MANUAL

OF THE

VERTEBRATES

OF THE

NORTHERN UNITED STATES,

INCLUDING

THE DISTRICT EAST OF THE MISSISSIPPI RIVER, AND NORTH OF NORTH CAROLINA AND TENNESSEE, EXCLUSIVE OF MARINE SPECIES.

BY

DAVID STARR JORDAN, M.S., M.D.

PROFESSOR OF NATURAL HISTORY IN N. W. C. UNIVERSITY, AND IN INDIANA STATE MEDICAL COLLEGE.

CHICAGO:

JANSEN, McCLURG & COMPANY.

1876.

THE LAKESIDE PRESS, CHICAGO.

PREFACE.

This book has been written to give collectors and students who are not specialists, a ready means of identifying the families, genera and species of our Vertebrate Animals. In deference to the uniform experience of botanists, and in view of the remarkable success achieved by Dr. COUES, in the application of the method to Ornithology, the author has adopted the system of artificial keys.

To keep the book of a size convenient for field use, class and ordinal characters have been abbreviated, since they do not lie directly within the purpose of the work; generic characters have been confined to the "key," under the head of each family, while for specific characters, only such points have been generally retained, as are distinctive as well as descriptive. The same necessity has led to the suppression of many of the doubtful or nominal "species," which still encumber our systematic works, and to the omission of synonymy and references to authorities, excepting in cases of recent or original changes of nomenclature.

Use has been freely made of every available source of information, and it is believed that the present state of our knowledge is fairly represented.

The account of the Mammals has been chiefly compiled from PROF. BAIRD'S "Mammals of North America," in the light of the recent revisions by Mr. J. A. ALLEN, Dr. COUES, and Prof. GILL. I am indebted to Mr. B. H. VAN VLECK, of Cambridge, Mass, for the revision of the proof-sheets of the Mammals and the Turtles.

I wish further to express my deep obligation to Dr. ELLIOTT COUES, having by his advice and consent, made free use of all his published writings. These have been drawn upon especially in the preparation of the keys to the Warblers, the Hawks, and other difficult groups, in the descriptions of the Water

Birds, and in the preliminary key to the orders of Birds. Dr. COUES has also kindly placed at my disposal the advance-sheets of his yet unpublished monograph of American Bats.

I am also under obligations to Messrs. BAIRD and RIDGWAY, whose writings have been to me of essential service; to Prof. COPE, whose numerous papers have been of great value in the systematic arrangement of the Fishes, and whose check list of the Reptiles and Batrachians has been closely followed in the classification of those groups; to Prof. GILL, whose arrangement of the families of Fishes has been, with few modifications, adopted in the present work; to Dr. GÜNTHER, whose Catalogue of Fishes is the vade-mecum of the working ichthyologist; to Prof. RICHARD BLISS, Jr., who has generously aided me in the identification of the Ichthelidæ, in advance of the publication of his monograph of that most difficult group; to Mr. E. W. NELSON, of Chicago, who has kindly furnished me with a yet unpublished list of the Birds and Reptiles of Illinois; to Prof. F. W. PUTNAM, Dr. B. G. WILDER, and Dr. C. C. ABBOTT, for the loan of books and other aids; to Prof. H. E. COPELAND, for help of various kinds, particularly in the verification of descriptions, and to the many collectors and compilers of local lists, who have freely placed their material at my disposal.

Our knowledge of the geographical distribution of many animals is still very imperfect. Quite a number of Southern and Western species, here excluded as extra-limital, will probably yet be found within the region included in this work.

Additions and corrections of every character are earnestly solicited from naturalists and teachers.

D. S. J.

INDIANAPOLIS, IND., May, 1876.

SIGNS, ABBREVIATIONS, ETC., EXPLAINED.

I.—Mammals.

i.=Incisor teeth.
c.=Canines.
pm.=Premolars.
m.=Molars.

Thus "i. $\frac{2}{1}\cdot\frac{2}{1}$" indicates two incisor teeth on each side in the upper jaw, and one on each side in the lower.

"Toes 5-4" implies fore feet five-toed, hind feet four-toed.

Other signs are as used in Birds.

II.—Birds.

L.=Length in inches (along back from tip of bill to end of longest tail feather); thus, "L. 7¼" means, length 7¼ inches.

E.=Extent (spread of wing) measured in inches.

W.=Length of wing (from bend of closed wing—carpal joint—to tip of longest feather) in inches.

T.=Length of tail in inches (*i.e.*, *actual* length of the longest tail feather.)

B.=Length of bill in inches (measured along culmen from tip to frontal feathers.)

Hd.=Length of head in inches (measured with dividers from base of bill to nape.)

Ts.=Length of tarsus in inches (measured in front.)

Tcl.=Length of middle toe with its claw.

The measurements given in the descriptions are understood to represent a fair *average adult male;* a variation of one-sixth, or more, in *absolute* length is nothing unusual; *relative* lengths, as of wings and tail, are much more constant. To save space I have preferred to say "L. 6," to saying "L. 5½ to 6½."

♂=Male.

♀=Female.

>=More than, longer than, or more than equivalent to.

<=Less than, in its various senses.

= =Sign of equivalence.

The toes are numbered 1, 2, 3, 4; 1 being the hind toe, or hallux; 2 the inner anterior toe; 3 the middle toe; and 4 the outer toe.

III.—Reptiles.

L.=Length in inches.

Sc. or *Scales*=Number of longitudinal rows of scales exclusive of the ventral series.

G.=Number of ventral plates, or gastrosteges, counted along the belly, from the throat to the vent. The figures given in the descriptions are intended to be *average*, the actual number being quite variable.

U.=Number of pairs of sub-caudal plates, or urosteges, counted from the vent to the tip of the tail.

IV.—Fishes.

D.=Dorsal fin.

2d D.=Second dorsal fin.

P.=Pectoral fins.

V.=Ventral fins.

A.=Anal fin.

C.=Caudal fin.

B.=Branchiostegals.

Roman numerals used with the above abbreviations indicate the number of *spines* or *inarticulate* rays in a fin. *Arabic* numerals indicate the number of *soft rays*. In a fin containing both spines and soft rays, a *comma* (,) separating the numerals indicates that the two kinds of rays are *continuous*, or more or less connected. A *dash* (—) indicates their *separation*. Thus, "D, X, 12," describes a single dorsal fin with 10 spines and 12 soft rays; "D, X—12," indicates two dorsal fins—the first of 10 spines, the second of 12 soft rays; "D, X—I, 12," would indicate the presence of a single spine in the second dorsal.

The posterior soft ray of the dorsal and anal fins is usually split to the base. It should be counted as *one* ray and not as *two*.

Lat. l.=Lateral line, *i. e.*, the number of scales contained in its course. When the lateral line is obsolete, "lat. l." signifies the number of scales in a row from the head to the base of the caudal fin. Thus, "lat. l. 36" means that there are 36 scales in a row along the sides from the head to the caudal.

In all cases the number of rays or scales, as given in the descriptions, is intended to represent a fair average, and a variation of one-sixth, or even more, in either direction need not surprise the student. Generally the spines and scales are more constant in their numbers than the soft rays, and the fewer of either, the less variable.

Depth in length=The greatest depth of the body as contained in the distance along the side from the snout to the base of the caudal.

Head in length=The distance from the snout along the cheeks to the extremity of the opercle, as contained in the distance from the snout to the base of the caudal.

Eye in head=Its longitudinal diameter as contained in the length of the side of the head.

As above stated, these measurements, as given in the descriptions, are *intended* to be the *average* of *living adults*, and must be applied to young specimens or preserved ones with caution.

Young fishes are always much more elongated than adults, and the eye is proportionally much larger.

A fin is said to be "*long*" when it has a long base, or is many-rayed. A "*high*" fin is one in which the individual rays are elongated.

PRINCIPAL ABBREVIATIONS OF NAMES OF AUTHORS CITED IN THIS VOLUME.

Ag.=Agassiz.
Allen=J. A. Allen.
Auct.=Auctorum (of Authors.)
Aud.=Audubon.
Aud. & Bach. = Audubon & Bachman.
Bd.=Baird.
B. & G.—Baird & Girard.
Bartr.—Bartram.
Beauv.=Palisot de Beauvais.
Bl. & Schn.=Bloch & Schneider.
Bodd.=Boddaert.
Bon., or *Bp.*=C. L. Bonaparte.
Brev.=Brevoort.
Brünn.=Brünnich.
Cab.=Cabanis.
Cass.=Cassin.
Coop.=Cooper.
Cuv.=Cuvier.
C. & V.=Cuvier and Valenciennes.
Daud.=Daudin.
Dek.=Dekay.
Desm.=Desmarest.
Dum.=Duméril.
Erxl.=Erxleben.
Fitz.=Fitzinger.
Forst.=Forster.
Grd.=Girard.
Gm., or *Gmel.*=Gmelin.
Gray (*Birds*)=G. R. Gray.
Gray (*Mam. & Rept.*)=J. E. Gray.
Gthr.=Günther.
Hald.=Haldeman.
Holbr.=Holbrook.
Illig.=Illiger.
Kirt.=Kirtland.
Lac.=Lacepède.
Lafr.=Lafresnaye.
Lath.=Latham.
Latr.=Latreille.
Lawr.=Lawrence.
LeC.=LeConte.
Less.=Lesson.
LeS.=LeSueur.
Licht.=Lichtenstein.
L.=Linnæus.
Mitch.=Mitchill.
Nutt.=Nuttall.
Pall.=Pallas.
Raf.=Rafinesque.
Reich.=Reichenbach.
Rich.=Richardson.
Ridg.=Ridgway.
Sab.=Sabine.
Sav.=Savigny.
Schl.=Schlegel.
Scl.=Sclater.
H. Smith=Hamilton Smith.
Steph.=Stephens.
Stor.=Storer.
Strickl.=Strickland.
Sund.=Sundevall.
Sw.=Swainson.
Temm.=Temminck.
Towns.=Townsend.
Val.=Valenciennes.
Vieill., or *V.*=Vieillot.
Vig.=Vigors.
Wagl.=Wagler.
Walb.=Walbaum.
Wils.=Wilson.

*** Names of Authors not in the above list are usually written in full.

VERTEBRATE ANIMALS

OF THE

NORTHERN UNITED STATES.

VERTEBRATA.

(*The Vertebrates.*)

The Vertebrates are, in popular language, "animals with a back-bone." They are distinguished from all other animals, says Prof. Huxley, "by the circumstance that a transverse and vertical section of the body exhibits two cavities, completely separated from one another by a partition. The dorsal cavity contains the cerebro-spinal nervous system; the ventral, the alimentary canal, the heart, and, usually, a double chain of ganglia, which passes under the name of the 'sympathetic.' A vertebrated animal may be devoid of articulated limbs, and it never possesses more than two pairs. These are always provided with an internal skeleton, to which the muscles moving the limbs are attached."

Without further discussion of this great group of animals, we will proceed at once to the consideration of the classes. Of existing vertebrates, we here recognize eight classes. These are: — 1. MAMMALIA, or Mammals; 2. AVES, or Birds; 3. REPTILIA, or Reptiles; 4. BATRACHIA, or Batrachians; 5. PISCES, or True Fishes; 6. ELASMOBRANCHII, or Selachians; 7. MARSIPOBRANCHII, or Myzonts; 8. LEPTOCARDII, or Lancelets.

Of these classes, two, *Elasmobranchii* and *Leptocardii*, are represented by marine forms only, and do not, therefore, come within the scope of this treatise.

The relations of the classes of Vertebrates may be recognized by the following analysis, taken, in part, from Prof. Gill's "Arrangement of the Families of Fishes." Only the most obvious characters are here referred to, although others, less striking, are often of greater taxonomic value.

CLASSES OF VERTEBRATES.

* Respiration never performed after birth by means of branchiæ.

† Exoskeleton developed as hair (rarely obsolete); warm blood; heart with four cavities; diaphragm complete; two occipital condyles; viviparous; young developed from a minute egg, and nourished for a time by milk secreted in the mammary glands of the mother. . . . . . MAMMALIA.

†† Exoskeleton developed as feathers; warm blood; heart with four cavities; diaphragm incomplete; a single occipital condyle; viviparous; young hatched from a comparatively large egg; no mammary glands. . . . . . AVES.

††† Exoskeleton developed as scales, or bony plates; cold blood; heart with three cavities (four in *Crocodilia*); a single occipital condyle; viviparous (or rarely ovoviparous); young hatched from a rather large egg. . . . REPTILIA.

** Respiration performed by gills for a part, or the whole, of life; cold blood.

‡ Skull more or less developed, with the notochord not continued forwards beyond the pituitary body; brain differentiated and distinctly developed; heart developed and divided into at least an auricle and ventricle.

a. Skull well developed and with a lower jaw; nostrils paired.

b. No median rayed fins; limbs not developed as rayed fins, but, if present, having the same skeletal elements as in the higher vertebrates; skin usually naked; respiration in the adult chiefly, or entirely, accomplished by means of lungs, the branchiæ rarely persistent. . . . . BATRACHIA.

bb. Rayed fins present on median line of the body; limbs developed as rayed fins; at least one pair being always present; respiration by means of branchiæ throughout life; lungs not developed.

c. Membrane bones (operculum, etc.) developed in connection with the skull; gills free; the branchial openings a single slit on each side; exoskeleton usually of scales, or plates, never placoid; eggs small and numerous. . . . . PISCES.

cc. Skull without membrane bones; gills not free; the branchial openings slit-like, usually several in number; exoskeleton placoid, never of scales, but usually composed of calcified papillæ of various styles; eggs few and large. . ELASMOBRANCHII.

aa. Skull imperfectly developed and with no lower jaw; a single median nostril; paired fins undeveloped, with no shoulder girdle nor pelvic elements; gills purse-shaped; skin naked. . . MARSIPOBRANCHII.

‡‡ Skull undeveloped, with the notochord persistent and extending to the anterior end of the head; brain not distinctly differentiated; heart none. LEPTOCARDII.

Class I.—Mammalia.

(*The Mammals.*)

A Mammal is a warm-blooded, air-breathing vertebrate, having the skin more or less covered with hair; viviparous; embryo developed from a minute egg, and provided with an amnion and allantois; young nourished for a time after birth by milk, secreted in the mammary glands of the mother; respiration never by branchiæ, but after birth by lungs, suspended freely in the thoracic cavity, which is completely separated from the abdominal cavity by a muscular septum (the diaphragm); heart with four cavities; a complete double circulation. The peculiarities of the skeleton are too numerous to be noticed in this connection.

The following key to the Orders and Families of Mammals which occur within our limits (omitting the Seals and the Cetaceans, all our members of which groups are marine), is mostly taken from Prof. Gill's "Arrangement of the Families of Mammals." It has been thought best not to give here any separate account of the different orders, as, in the abundance of literature on that subject, it seems unnecessary.

ORDERS OF MAMMALIA.

* Young not born until of considerable size and nearly perfect development, deriving its nourishment, before birth, from the mother through the intervention of a placenta; a well developed corpus callosum. (MONODELPHIA.)

† Brain with a relatively large cerebrum, overlapping much, or all, of the cerebellum and olfactory lobes. (EDUCABILIA.)

‡ Anterior and posterior limbs and pelvis well developed; femur and humerus not exserted beyond the common integuments of the body.

a. Feet with developed claws; canines specialized; molars, one or more, sectorial, adapted for cutting; incisors $\frac{6}{6}$; chiefly carnivorous. . . CARNIVORA, A.

aa. Feet with hoofs; molars mostly with grinding surfaces; incisors various; no tusks; developed toes, four or fewer; chiefly herbivorous. . . UNGULATA, B.

†† Brain with a relatively small cerebrum, leaving behind much of the cerebellum exposed, and, in front, much of the olfactory lobes. (INEDUCABILIA.)

b. Canine teeth present, in some form; incisors not $\frac{2}{2}$ nor $\frac{4}{2}$.

c. Anterior limbs adapted for flight; ulna and radius united; bones of hand and fingers much elongated, supporting a thin, leathery skin, extending along sides of body to the posterior limbs; mammæ pectoral. . . CHIROPTERA, C.

cc. Limbs all adapted for walking; ulna and radius not united; hand normal; mammæ usually abdominal. . . . . . INSECTIVORA, D.

bb. No canine teeth; incisors $\frac{2}{2}$, rarely $\frac{4}{2}$, chisel shaped; limbs adapted for walking. . RODENTIA, E.

** Young born when of very small size and incomplete development, never connected by a placenta to the mother; corpus callosum rudimentary. (DIDELPHIA.) . MARSUPIALIA, F.

A. FAMILIES OF CARNIVORA.

* Intestinal canal provided with a cœcum; feet digitigrade; toes 5–4.

† Teeth 28 to 30; head broad; snout short; claws sharp, retractile. . . . . FELIDÆ, 1.

†† Teeth 38 to 46; snout elongated; claws bluntish, nonretractile. . . . . CANIDÆ, 2.

** Intestinal canal without a cœcum; feet plantigrade, or, if not, toes 5–5.

‡ Teeth less than 40; body rather slender; feet often more or less perfectly digitigrade. . . . MUSTELIDÆ, 3.

‡‡ Teeth 40 or 42; body stout; feet completely plantigrade.

a. Tail rudimentary; teeth 42; lower jaw with three true molars; body very large and heavy. . URSIDÆ, 4.

aa. Tail well developed; teeth 40; lower jaw slender, with two true molars; body moderately stout. PROCYONIDÆ, 5.

B. FAMILIES OF UNGULATA.

* Toes paired (artiodactylous); horns solid, deciduous; teeth 34 or 32. CERVIDÆ, 6.

C. FAMILIES OF CHIROPTERA.

* Snout not appendaged; tail inclosed in the membrane, or only the last joint exserted. . . . VESPERTILIONIDÆ, 7.

D. FAMILIES OF INSECTIVORA.

* No external ear; fore feet greatly enlarged—adapted for digging; limbs very short. . . . TALPIDÆ, 8.

** An external ear; feet normal; hind feet usually longest. SORICIDÆ, 9.

E. FAMILIES OF RODENTIA.

* Incisors $\frac{2}{2}$; tail well developed.

† Hair forming a fur, without spines.

a. Tail broad, flat and scaly; feet webbed; molars $\frac{4 \cdot 4}{5 \cdot 5}$; body stout and heavy. . . . CASTORIDÆ, 11.

aa. Limbs very short, about equal; fore claws much enlarged; large external cheek pouches; body thick-set and heavy; molars $\frac{4 \cdot 4}{4 \cdot 4}$. . . . . . GEOMYIDÆ, 12.

aaa. Tail and hind legs excessively elongated, the latter adapted for leaping; molars $\frac{4 \cdot 4}{3 \cdot 3}$. . ZAPODIDÆ, 13.

aaaa. With none of the preceding combinations.

b. Molars $\frac{5 \cdot 5}{4 \cdot 4}$ or $\frac{4 \cdot 4}{4 \cdot 4}$; no foramen (as in the next); tail usually long and hairy; (squirrels, etc.) SCIURIDÆ, 10.

bb. Molars $\frac{2 \cdot 2}{2 \cdot 2}$ to $\frac{4 \cdot 4}{3 \cdot 3}$; zygomatic process of upper jaw with a foramen; (rats and mice). . . . MURIDÆ, 14.

†† Fur with stiff spine-like bristles; toes with long, curved claws. . . . . SPALACOPODIDÆ, 15.

** Incisors $\frac{4}{2}$; the middle upper incisors large, grooved, the outer ones small; teeth 28; tail very short; ears long. LEPORIDÆ, 16.

F. FAMILIES OF MARSUPIALIA.

* Tail long, prehensile, nearly naked; feet plantigrade; incisors $\frac{5 \cdot 5}{4 \cdot 4}$; teeth 50. . . . . . . DIDELPHIDIDÆ, 17

ORDER A.—CARNIVORA.

(The Flesh-Eaters or Feræ.)

FAMILY I.—FELIDÆ.

(The Cats.)

Digitigrade Carnivora with the toes 5–4; claws compressed, very sharp, retractile; palms and soles hairy, with naked pads under each toe and the ball of the foot. Body compact; head short, broad and rounded. Dentition i. $\frac{3 \cdot 3}{3 \cdot 3}$; c. $\frac{1 \cdot 1}{1 \cdot 1}$; pm. $\frac{3 \cdot 3}{2 \cdot 2}$ or $\frac{2 \cdot 2}{2 \cdot 2}$; m. $\frac{1 \cdot 1}{1 \cdot 1}$=30 or 28; canine teeth long and sharp; teeth all strongly trenchant; tongue with short, retrorse papillæ. General aspect cat-like. A well-known group comprising about fifty species, found in all parts of the world excepting Australia and its islands.

* Premolars $\frac{3 \cdot 3}{2 \cdot 2}$, anterior upper one very small; tail at least half as long as the body (exclusive of head and neck); fur compact and glossy; ears not tufted. . . . . . . FELIS, 1.

** Premolars $\frac{2 \cdot 2}{2 \cdot 2}$, (anterior upper one wanting); tail less than half length of body proper; ears triangular, tufted. LYNX, 2.

1. FELIS, Linnæus. Cats.

1. ***F. concolor,*** L. American Panther. Cougar. Puma. Above tawny brownish yellow; a wash of darker along dorsal line; dirty white below; large; body $4\frac{1}{2}$ feet long. Whole continent of America, N. to lat. 50° to 60°.

2. LYNX, Rafinesque. Wild Cats.

1. ***L. canadensis,*** (Desm.) Raf. Canada Lynx. Feet very large, densely furred beneath in winter, concealing the small, naked patches; tail black at tip; no distinct bars on inner side of legs; much larger than the next, with larger feet and longer fur. N. A.

2. ***L. rufus,*** (Guldenstädt.) Raf. American Wild Cat. Inner sides of legs with dark cross bands; tail with a black patch at end above, preceded by half rings. U. S. and northward.

FAMILY II.—CANIDÆ.

(*The Dogs.*)

Digitigrade Carnivora with blunt, non-retractile claws; toes 5–4. Muzzle more or less elongated. Dentition typically i. $\frac{3\cdot3}{3\cdot3}$; c. $\frac{1\cdot1}{1\cdot1}$; pm. $\frac{4\cdot4}{4\cdot4}$; m. $\frac{2\cdot2}{3\cdot3}=42$; canines large, rather blunt. Both hemispheres.

* Tail comparatively short; upper incisors distinctly lobed; pupil circular. . . . . Canis, 1.

** Tail comparatively long and bushy; upper incisors scarcely lobed; pupil elliptical; body more slender.

† Tail with soft fur and long hair; muzzle long. . Vulpes, 2.

†† Tail with a concealed mane of stiff hairs, and without soft fur; muzzle shorter. . . . . . Urocyon, 3.

1. ***CANIS,*** Linnæus. Wolves.

1. ***C. lupus,*** L. Wolf. Color exceedingly variable; northward it is chiefly gray, southward more and more blackish and reddish, till in Florida black wolves predominate, and in Texas red ones. N. A. and northern parts of the Old World. (*C. occidentalis*, Auct.)

2. ***VULPES,*** Brisson. Foxes.

1. ***V. vulgaris,*** Fleming. Red Fox. Cross Fox. Silver Fox. Black Fox. Chiefly red, with black feet and ears; tip of tail white. A single species, widely variable in color, as indicated by the common names. N. Am. Europe (*V. fulvus*, Auct.)

3. ***UROCYON,*** Baird. Gray Foxes.

1. ***U. virginianus,*** (Erxleben.) Gray Fox. Chiefly gray; fur dusky or fulvous, hairs hoary at tip; tip of tail usually dark. Penn. S., W. to the Pacific.

FAMILY III.—MUSTELIDÆ.

(The Weasels.)

Carnivora either plantigrade or digitigrade, with the toes 5–5. Molars $\frac{1\cdot1}{2\cdot2}$ (rarely $\frac{1\cdot1}{1\cdot1}$); the upper and the last lower one tubercular; no cœcum. Most species provided with glands near the anus which secrete a fetid liquid. Some are strictly carnivorous while others are rather omnivorous. Size usually medium or small. They are found in all parts of the earth excepting the Australian region.

* Last or true molar of upper jaw short, small, transversely elongated; toes short; claws retractile. (Mustelinæ.)

† Teeth 38; body slender; feet digitigrade; tail rather long; lower first molar with an internal tubercle. . Mustela, 1.

†† Teeth 34; body slender; feet digitigrade; tail usually long; sectorial tooth without internal tubercle. . PUTORIUS, 2.

††† Teeth 38; body stout; feet sub-plantigrade; tail as long as head, very full and bushy. . . . . . GULO, 3.

** True molar of upper jaw very large, sub-triangular, tubercular; fore claws much lengthened, for digging. (MELINÆ.)

a. Body short and stout; tail very short; teeth 32. TAXIDEA, 4.

*** True molar of upper jaw quadrangular, wide, very large, with an outer cutting ridge; claws non-retractile, fore claws lengthened, fossorial. (MEPHITINÆ.)

b. Snout pointed; colors black and white; fetid glands highly developed; teeth 34. . . . . MEPHITIS, 5.

**** True molar large, quadrate; teeth 36; body elongated; toes palmate, adapted for swimming. (LUTRINÆ.) LUTRA, 6.

1. MUSTELA, Linnæus. MARTENS.

1. ***M. martes,*** L. SABLE. PINE MARTEN. Chiefly reddish yellow, clouded with black; legs and tail, blackish; length less than two feet. Maine to Oregon, and northward; also in Old World. (*M. americana*, Turton.)

2. ***M. pennantii,*** Erxleben. FISHER. BLACK CAT. Color, variable; belly, legs, tail, and hind parts chiefly black; length more than two feet. Northern N. A.

2. PUTORIUS, Cuvier. WEASELS.

1. ***P. vulgaris,*** Cuvier. LEAST WEASEL. Liver-brown, white beneath; usually white in winter; tail never really black at tip; body 6 to 7 inches long. N. U. S. and northward, also Old World.

2. ***P. ermineus,*** Cuvier. COMMON WEASEL. ERMINE. Colors various; tip of tail usually black; white in winter, northward. Length of body 8 to 11 inches.

N. Am. and Old World, abundant. (*P. noveboracensis*, *cicognanii*, etc., of authors.)

3. ***P. lutreolus,*** Cuvier. COMMON MINK. Brownish chestnut; tail black or nearly so; length of body 15 to 20 inches. N. A. and Old World. (*P. vison*, Gapper.)

3. GULO, Storr. WOLVERINES.

1. ***G. luscus,*** (L.) Sabine. WOLVERINE. Dark brown; legs and beneath black. N. U. S. and N.

4. TAXIDEA, Waterhouse. BADGERS.

1. ***T. americana,*** (Bodd.) Baird. AMERICAN BADGER. Chiefly grayish. Wisconsin, N. and W.

5. MEPHITIS, Cuvier. SKUNKS.

1. ***M. mephitica,*** (Shaw) Baird. COMMON SKUNK. Usually black with tip of tail, dorsal stripes and nuchal patch white; sometimes all black or even nearly all white. Mexico to Arctic regions, an abundant and familiar species. (*M. chinga*, Tiedemann.)

6. LUTRA, Linnæus. OTTERS.

1. ***L. canadensis,*** Sabine. AMERICAN OTTER. Liver-brown; length 4½ feet. U. S.

FAMILY IV.—URSIDÆ.

(*The Bears.*)

Plantigrade Carnivora having the body thick and clumsy. Tail rudimentary. Teeth 42; molars broad and tuberculated, according with the omnivorous diet. Species few and widely distributed,—in North America, there are probably but two, although many have been described; these are, as has been shown by Mr. J. A. Allen, the

Polar Bear (*U. maritimus*), and the following which is the common Bear of the Old World.

1. URSUS, Linnæus. Bears.

1. ***U. arctos,*** L. Brown, Black, Cinnamon and Grizzly Bear. Color, size, etc., exceedingly variable, but the several forms or varieties intergrade perfectly. N. Am. and Europe. (*U. americanus*, Pallas. *U. horribilis*, Ord.)

FAMILY V.—PROCYONIDÆ.

(*The Raccoons.*)

Plantigrade Carnivora of moderate size, with the body comparatively slender and the tail well developed. Teeth 40. Snout more or less elongated; no cœcum. Genera two,—*Nasua*, the Coatimundi of Mexico, and the following, all American.

1. PROCYON, Storr. Raccoon.

1. ***P. lotor,*** (L.) Storr. Common Raccoon. "Coon." Grayish white; hairs black-tipped; tail with black rings. U. S.; abundant.

ORDER B.—UNGULATA.

(*The Hoofed Mammals*).

FAMILY VI.—CERVIDÆ.

(*The Deer*).

Horns deciduous, solid, developed from the frontal bone, more or less branched, covered at first by a soft, hairy integument, known as "velvet"; when the horns attain their full size, which they do in a very short time, there arises at the base of each a ring of tubercles known as the "burr;" this compresses and finally obliterates the blood-vessels supplying the velvet, which dries up and is stripped off, leaving the bone hard and insensible; the horns or "antlers" are shed annually, the separation of the "beam" from its "pedicel" taking place just below the burr ; the antlers are wanting in the female (excepting in the Reindeer) but they are usually present in the male. Herbivorous; stomach in four divisions, of the ordinary ruminant pattern. Dental formula, i. $\frac{0\text{-}0}{3\text{-}3}$; c. (usually) $\frac{0\text{-}0}{1\text{-}1}$; pm. $\frac{3\text{-}3}{3\text{-}3}$; m. $\frac{3\text{-}3}{3\text{-}3}$. A widelv distributed family.

* Horns (in males only) very broadly palmated to the tip; nose very broad, entirely hairy except a small naked spot between nostrils. . . . . ALCE, 1.

** Horns (in both sexes) broadly palmated at tip; nose entirely hairy. . . . . RANGIFER, 2.

*** Horns (in males only) rounded more or less; rarely sub-palmated; nose naked and moist.

† Horns large, curving backward, with the snags all directed forward, one of them immediately above the burr; tail very short; hoofs broad and rounded; size very large. CERVUS, 3.

†† Horns smaller, curving forward, the first snag short, at some distance above the base, and like the others curving upward; tail rather long; hoofs rather elongate; size smaller. . . . . CARIACUS, 4.

1. ALCE, Hamilton Smith. ELKS.

1. ***A. americanus,*** Jardine. MOOSE. AMERICAN ELK. Largest of our *Cervidæ*, reaching the size of a horse. Maine and N. N. Y. to Oregon and N.

2. RANGIFER, Hamilton Smith. REINDEER.

1. ***R. caribou,*** Aud. & Bach. WOODLAND CARIBOU. REINDEER. Maine to Hudson's Bay and Lake Superior; perhaps the same as the Reindeer of Europe. (*R. tarandus.*)

3. ***CERVUS,*** Linnæus. DEER.

1\. ***C. canadensis,*** Erxleben. "AMERICAN ELK." WAPITI. Chestnut red, grayish in winter; size nearly equal to that of the Moose. Alleghany region of Penn. and Va., Wis. (near Green Bay, *Van Vleck*,) Yellowstone region and W.

4. ***CARIACUS,*** Gray. RED DEER.

1\. ***C. virginianus,*** (Bodd.) Gray. VIRGINIA DEER. RED DEER. General color chestnut red, grayish in winter. Maine to Rocky Mountains and S., abundant in many regions.

ORDER C.—CHIROPTERA.

(The Bats.)

FAMILY VII.—VESPERTILIONIDÆ.

(The Ordinary Bats.)

Insectivorous Bats with the snout not appendaged, or merely with two lateral excrescences. Wing membranes ample. Tail completely enclosed in the interfemoral membrane or only the last joint exserted.

* Cheeks without excrescences; ears moderate.

† Incisors $\frac{2}{3}\cdot\frac{2}{3}$. . . . . . . VESPERTILIO, 1.

†† Incisors $\frac{1}{3}\cdot\frac{1}{3}$. . . . . . . ATALAPHA, 2.

** Cheeks with two large excrescences; ears excessively large, an inch high; teeth 36; i. $\frac{2}{3}\cdot\frac{2}{3}$. . . CORYNORHINUS, 3.

1. ***VESPERTILIO,*** Linnæus. TYPICAL BATS.

>*Scotophilus*, Leach.

* Teeth 38; molars $\frac{6}{6}\cdot\frac{6}{6}$; species with thin wings and ears. (*Vespertilio.*)

1\. ***V. subulatus,*** Say. LITTLE BROWN BAT. Face

small, fox-like, with high forehead and pointed snout; ears large, oval, twice the height of the erect tragus; wings naked; interfemoral membrane naked except at base, face whiskered; color dull olive-brown. L. 3; E. 9; T. 1½. N. Am.; abundant every where. A small and very variable species. (*V. lucifugus*, Le C.)

** Teeth 36; molars $\frac{5}{6}\cdot\frac{5}{6}$. (*Vesperides.*)

2. ***V. noctivagans,*** LeC. Silver Black Bat. Tragus almost as broad as high, scarcely one-third height of ear; femoral membrane entirely though scantily furred; fur long and silky, black, usually with silvery tips to the hairs. L. 3½; E. 12; T. 1½. U. S. generally.

*** Teeth 34; molars $\frac{5}{5}\cdot\frac{5}{5}$; stouter species, with thicker wings and more leathery ears. (*Vesperugo.*)

3. ***V. georgianus,*** F. Cuvier. Georgia Bat. Tragus slender, erect, half the height of the auricle; upper incisors about equal in size; femoral membrane one-third furred; dark reddish brown, brighter forwards. L. 3; E. 9; T. 1½. Maine to Texas; chiefly southward.

**** Teeth 32; molars $\frac{4}{5}\cdot\frac{4}{5}$; stout species. (*Vesperus.*)

4. ***V. fuscus,*** Beauv. Carolina Bat. Dusky Bat. Tragus never pointed, nearly half as high as auricle; wings naked; interfemoral membrane furred at base; ears more or less turned outward; upper lateral incisors small, scarcely visible. L. 3 to 4; E. 12; T. 1½. U. S.; a widely diffused species. (*V. carolinensis*, Geoff. St. Hilaire.)

2. *ATALAPHA,* Rafinesque. Red Bats.

* Teeth 30; molars $\frac{4}{5}\cdot\frac{4}{5}$; upper incisors small; wings naked and interfemoral membranes nearly so. (*Nycticejus.*)

1. ***A. crepuscularis,*** (Le C.) Coues. Twilight Bat. Ears small, wide apart; a small wart above eye; fur

rather scanty; dark fawn color above, passing into brownish below; L. $3\frac{1}{3}$; E. 9; T. $1\frac{1}{3}$. Penn. to Mo., and S.W.

** Teeth 32; molars $\frac{5}{5}\cdot\frac{5}{5}$; upper incisors stout; interfemoral membrane hairy above, and wings with furry patches (*Lasiurus*.)

2. ***A. noveboracensis,*** (Erxl.) Coues. RED BAT. Fur long and silky, reddish brown, mostly white at tip; lips and ears not edged with black; a whitish tuft at base of thumb; L. $3\frac{3}{4}$; E. 12; T. $1\frac{3}{4}$. U. S., every where. An abundant species, readily known by its reddish color.

3. ***A. cinereus,*** (Beauv.) Coues. HOARY BAT. Rich chocolate-brown, overlaid with white; lips and ears marked with black; L. 5; E. 14; T. $2\frac{1}{4}$. U. S., rather northward, rare. (*V. pruinosus*, Say.)

3. CORYNORHINUS, Harrison Allen. BIG-EARED BATS.

1. ***C. macrotis,*** (Le C.) H. Allen. BIG-EARED BAT. Blackish, fur soft and long; L. $3\frac{1}{2}$; E. 11; T. $1\frac{3}{4}$. Va. to Missouri region and S.

ORDER D.—INSECTIVORA.

(*The Insect Eaters.*)

FAMILY VIII.—TALPIDÆ.

(*The Moles.*)

Body stout, thick, and clumsy, without visible neck. Eyes rudimentary, sometimes concealed. No external ears. Limbs very short; feet greatly expanded and provided with strong claws, adapted for digging; anterior limbs much larger than posterior. Scapula as long as humerus and radius together. Canines usually present.

Fur compact, soft and velvety. Found on both continents.

* Nose elongated, not fringed; nostrils superior or lateral; tail short.

† Teeth 36; nostrils partly superior; tail nearly naked. SCALOPS, 1.

†† Teeth 44; nostrils lateral; tail densely hairy. SCAPANUS, 2.

** Nose elongated, fringed at end with a circle of long fleshy projections; nostrils terminal; tail nearly as long as body; teeth 44. . . . . CONDYLURA, 3.

1. SCALOPS, Cuvier. SHREW MOLES.

1. ***S. aquaticus,*** (L.) Fischer. COMMON MOLE. Dark plumbeous, paler below; feet full webbed; eyes not covered by skin. E. U. S.; an abundant and well known species.

2. ***S. argentatus,*** Aud. & Bach. PRAIRIE MOLE. Silvery plumbeous; said to be larger and more silvery than the preceding. Ohio and W., chiefly in the prairie region.

2. SCAPANUS, Pomel. HAIRY-TAILED MOLES.

1. ***S. breweri,*** (Bach.) Pomel. HAIRY-TAILED MOLE. Dark plumbeous, with brown gloss. E. U. S.

3. CONDYLURA, Illiger. STAR-NOSED MOLES.

1. ***C. cristata,*** (L.) Desmarest. STAR-NOSED MOLE. Blackish. U. S., chiefly northerly, from the Atlantic to the Pacific.

FAMILY IX.—SORICIDÆ.

(The Shrews.)

Mouse-like Insectivora, with the eyes and external ears distinctly developed. Muzzle elongated. Feet normal, not fossorial; the fore-feet mostly smaller than the hind ones. Teeth $\frac{16 \text{ to } 20}{12}$, no canines. The most abundant and widely distributed family of the Insectivora. The assumed specific distinctions in this family are utterly worthless, and it is at present impossible to characterize the species. The following are the best known.

* Ears large; auricle directed backwards; tail about as long as the body; upper anterior incisors with a second basal hook, and an angular process on the inner side near the point. SOREX, 1.

** Ears small, concealed by the fur; tail not longer than head; auricle directed forwards; upper anterior incisor without above-mentioned hook and process. . . BLARINA, 2.

1. SOREX, Linnæus. SHREWS.

1. ***S. platyrhinus,*** (Dek.) Wagner. COMMON SHREW. A small, long-nosed, large-eared species, of a chestnut color, having the tail much shorter than head and body; said to be abundant in the Eastern and Middle States.

2. ***S. cooperi,*** Bach. WESTERN SHREW. Still smaller; color light chestnut; tail nearly as long as head and body. New England, N. and W.

2. BLARINA, Gray. SHORT-TAILED SHREWS.

1. ***B. brevicauda,*** (Say) Baird. MOLE SHREW. Tail less than one-third length of head and body; color brownish gray. E. U. S., the common species.

ORDER E. — RODENTIA.

(*The Rodents or Glires.*)

FAMILY X.—SCIURIDÆ.

(*The Squirrels.*)

Rodents with the molars $\frac{5\cdot5}{4\cdot4}$ (upper anterior often deciduous), the last 4 of nearly equal size; a distinct postorbital process of frontal bone; tibia and fibula distinct. Species of rather small size, widely distributed. The variations in color, etc., are extremely great, and the number of well-defined species is very much less than was once supposed.

* A densely furred membrane extending along the sides between the anterior and posterior limbs; tail depressed, flattened, densely furred; permanent molars 5–5 above. SCIUROPTERUS, 1.

** Sides without membrane for "flying."

† No cheek pouches; tail bushy, at least as long as body; ears long; no black stripes along the back. . SCIURUS, 2.

†† Cheek pouches present.

‡ Back with 3 to 5 distinct black stripes; tail shorter than body, not bushy; cheek pouches well developed. TAMIAS, 3.

‡‡ Not as above; body rather slender, squirrel-like; cheek pouches well developed; claw of thumb rudimentary or wanting. . . . . SPERMOPHILUS, 4.

‡‡‡ Body large, thick-set, depressed; cheek pouches shallow; thumb rudimentary, armed with a small flat nail, not a claw; soles naked. . . . ARCTOMYS, 5.

1. ***SCIUROPTERUS,*** F. Cuvier. FLYING SQUIRRELS.

< *Pteromys*, Cuvier.

1. ***S. volans*** (L.) Coues. COMMON FLYING SQUIRREL. Yellowish brown, creamy white below. N. Am., abundant. (*P. volucella*, etc., authors.)

2. ***SCIURUS,*** Linnæus. SQUIRRELS.

1. ***S. cinereus,*** Linn. FOX SQUIRREL. Upper molars permanently 4; tail very large and bushy; general color rusty gray, varying from almost white, through various shades of rusty red to jet black, the latter color rare northward, reddish and orange shades predominating westward; L. 26; T. 14. E. U. S., very abundant in the

Mississippi region. Leading varieties are: var. *niger*, the Southern form, gray to black with the ears and nose white; var. *cinereus*, the Eastern form, with short ears, scarcely longer than the fur; and var. *ludovicianus*, the Western form, with high ears and a prevailing tinge of orange red.

2. ***S. carolinensis,*** Auct. GRAY SQUIRREL. BLACK SQUIRREL. Upper molars 5; varies from grizzly yellowish gray to pure jet black; L. 24; T. 13. N. Am., E. of the plains, abundant.

3. ***S. hudsonius,*** Pallas. RED SQUIRREL. CHICKAREE. Chiefly rusty red, back with a wash of brighter red; L. 14; T. 6½. N. Am., rather northerly; abundant.

3. *TAMIAS,* Illiger. GROUND SQUIRRELS.

1. ***T. striatus,*** (L.) Baird. CHIPMUNK. GROUND SQUIRREL. Back and sides with five black stripes; L. 11; T. 4½. Maine to Va., and westward; abundant.

4. *SPERMOPHILUS,* Cuvier. GOPHERS.

1. ***S. tridecemlineatus,*** (Mitch.) Aud. & Bach. STRIPED GOPHER. Dark brown, with light stripes alternating with lines of dots, about thirteen in all; L. 10; T. 4. Prairies; Ark. to the Saskatchawan.

2. ***S. franklini,*** (Sab.) Rich. GRAY GOPHER. Yellowish gray, mottled with brown; L. 15; T. 5½. Prairies; Ill. and northward.

5. *ARCTOMYS,* Schreber. MARMOTS.

1. ***A. monax,*** (L.) Gmel. WOODCHUCK. GROUND HOG. Grizzly gray, varying to chestnut and blackish: Hudson's Bay to Va., and westward; common.

FAMILY XI.—CASTORIDÆ.

(*The Beavers.*)

Aquatic Rodents of large size, having the molars rootless, $\frac{4\cdot4}{4\cdot4}$, or $\frac{5\cdot5}{4\cdot4}$; fore feet with five distinct toes and claws; tibia and fibula distinct; no postorbital process. A small family, containing but two existing genera, *Anisonyx* of our N.W. coast, and *Castor*.

* Molars 4-4 above; hind feet webbed; tail broad, flat, and scaly. CASTOR, 1.

1. CASTOR, Linnæus. BEAVERS.

1. ***C. fiber,*** L. BEAVER. Mexico to the Polar Sea; now being rapidly exterminated. (*C. canadensis*, Kuhl.)

FAMILY XII.—GEOMYIDÆ.

(*The Pouched Gophers.*)

Cheek pouches large and distinct, opening outside of the mouth. Molars $\frac{4\cdot4}{4\cdot4}$; incisors large and thick; skull heavy; temporal bones enormously developed. Limbs about equal, the fore claws, five in number, very large; tibia and fibula united. Body thick-set and clumsy.

Genera two, *Geomys* and *Thomomys*, all North American, and chiefly inhabiting the central plains; habits nocturnal and subterranean.

* A large groove near the middle of each upper incisor; crown of molars elliptical. . . . . . GEOMYS, 1.

1. GEOMYS, Rafinesque. POUCHED GOPHERS.

1. ***G. bursarius,*** (Shaw) Rich. POCKET GOPHER. Reddish brown, with plumbeous tinge. Prairies, Wis. and Ills., westward.

FAMILY XIII.—ZAPODIDÆ.

(*The Jumping Mice.*)

Hind legs greatly elongated, adapted for taking long leaps; fore legs short. Tail very long. Molars $\frac{4\cdot4}{3\cdot3}$;

tibia and fibula united. Contains, probably, but a single species, inhabiting the Northern U. S., related to the Old World Jerboas.

1. **ZAPUS,** Coues JUMPING MICE.

= *Jaculus*, Wagler.

1. ***Z. hudsonius,*** (Zimmermann) Coues. Yellowish brown. U. S., chiefly northward; variable.

FAMILY XIV.—MURIDÆ.

(*The Mice.*)

Incisors $\frac{2}{2}$; molars usually $\frac{3\cdot3}{3\cdot3}$; anteorbital foramen a vertical slit, widening above and bounded externally by a broad plate of the upper maxillary; coronoid and condyloid processes of lower jaw well developed. A large family, found in all parts of the globe, some of the species (*Mus*) being cosmopolitan, having accompanied man in his migrations through the ages; all are of small size, the muskrat being one of the largest, and many are smaller than any other quadrupeds, except the Shrews.

* Incisors laterally compressed; molars rooted. (MURINÆ.)

† Molars with three tubercles in transverse series; soles naked; tail long, scaly; (Old World species, introduced in America.) MUS, 1.

†† Molars complicated, with two tubercles in transverse series; soles usually hairy; (New World species.)

a. Rat-like; size large; ears large, rarely naked. NEOTOMA, 2.

aa. Mouse-like; size small.

b. Incisors not grooved; ears very large; feet and lower parts usually white. . . . HESPEROMYS, 3.

bb. Size very small; upper incisors grooved longitudinally along their anterior surface. . OCHETODON, 4.

** Incisors as broad as deep; molars rootless (except in *Evotomys*).

‡ Tail not specially compressed; teeth not grooved.

c. Molars rooted; coronoid process of lower jaw, not reaching level of condyle; ears overtopping the fur. EVOTOMYS, 5.

cc. Molars rootless; coronoid process as high as condyle; ears more or less concealed. . . . ARVICOLA, 6.

‡‡ Upper incisors grooved; root of lower incisor ending abruptly opposite the last molar; ears large. SYNAPTOMYS, 7.

‡‡‡ Tail scant-haired, vertically flattened; hind feet partly webbed; size large. . . . . . FIBER, 8.

1. ***MUS,*** Linnæus. OLD WORLD RATS AND MICE.

1. ***M. decumanus,*** Pallas. BROWN OR NORWAY RAT. Tail nearly an inch shorter than head and body, grayish brown above; paler below; feet dusty white; fur mixed with stiff hairs; cosmopolitan; introduced into America about 1775, and now the commonest species.

2. ***M. rattus,*** L. BLACK RAT. Tail not shorter than head and body; sooty black, plumbeous below; feet brown; introduced about 1544, but now being supplanted by the preceding.

3. ***M. alexandrinus,*** Geoff. ROOF RAT. WHITE-BELLIED RAT. Introduced in the Southern States.

4. ***M. musculus,*** L. COMMON HOUSE MOUSE. Cosmopolitan; every where too well known.

2. ***NEOTOMA,*** Say. & Ord. WOOD RATS.

1. ***N. floridana,*** S. & O. WOOD RAT. Tail scantily hairy, scarcely as long as body without head; feet entirely white; L. 14; T. 6, or less. S. U. S., N. to Mass. and Ills.

3. ***HESPEROMYS,*** Waterhouse. WHITE-FOOTED MICE.

* Fur soft and glossy; lower parts white; soles naked, or slightly hairy; tail closely hairy; ears large. (*Vesperimus.*)

1. ***H. leucopus,*** (Raf.) LeC. DEER MOUSE. WHITE-

FOOTED MOUSE. Yellowish brown; tail distinctly bicolor, about as long as head and body; hind feet more than $\frac{3}{4}$ inch. N. Am.; abundant.

2. ***H. michiganensis,*** (Aud. & Bach.) Wagner. MICHIGAN MOUSE. Tail little longer than body without head, bicolor; hind feet less than $\frac{3}{4}$ inch; dark brown; a darker dorsal band; L. 3, or less, to base of tail. Upper Miss. Valley to Mich., etc.

3. ***H. aureolus,*** (Aud. & Bach.) Wagner. RED MOUSE. Yellowish cinnamon, bright especially on ears; belly not pure white. Pa. to Ills., and S.

** Soles naked; tail scant-haired. about as long as head and body; ears small. (*Oryzomys.*)

4. ***H. palustris,*** (Harlan) Baird. RICE-FIELD MOUSE. Blackish and ashy above, becoming paler below; fur harsh, but compact; a large rat-like species. S. States, N. to N. J. and Kas.

4. OCHETODON, Coues. HARVEST MICE.

< *Reithrodon*, Baird.

1. ***O. humilis,*** (Aud. & Bach.) Coues. HARVEST MOUSE. Tail shorter than head and body; appearance decidedly mouse-like. U. S., southerly, N. to Iowa.

5. EVOTOMYS, Coues. LONG-EARED MICE.

< *Arvicola*, Bd.

1. ***E. rutilus*** (Pall.) var. ***gapperi,*** (Vigors) Coues. LONG-EARED MOUSE. Color chestnut; ears prominent; a brownish dorsal band; size of common mouse. Northern frontier, S. to Mass.

6. ARVICOLA, Lacepede. FIELD MICE.

* Back upper molar with two external triangles and a posterior crescent; middle upper molar with two internal triangles; front lower molar with three internal and two or three lateral triangles; size large. (*Myonomes.*)

1. ***A. riparius,*** Ord. MEADOW MOUSE. Fore claws not longer than hind claws; tail one-third length of head and body, or more. U. S.; generally abundant.

** Back upper molar with one exterior triangle and a posterior trefoil; middle upper molar with one internal triangle; front lower molar with two internal and one external triangle; fore claws not larger than hinder; fur ordinary; size medium. (*Pedomys.*)

2. ***A. austerus,*** LeC. Tail one-third length of head and body, or less. Western States, E. to Michigan.

*** Teeth as in *Pedomys;* fore claws larger than hinder; fur dense, silky, mole-like; size small. (*Pitymys.*)

3. ***A. pinetorum,*** LeC. PINE MOUSE. Tail about one-fourth length of head and body. E. U. S.

7. *SYNAPTOMYS,* Baird. COOPER'S MOUSE.

1. ***S. cooperi,*** Baird. Head short and heavy; fur soft and long. W. States, E. to Ind.

8. *FIBER,* Cuvier. MUSKRATS.

1. ***F. zibethicus,*** (L.) Cuv. MUSKRAT. MUSQUASH. A well-known aquatic animal, the largest of our *Muridæ.* N. Am.; every where.

FAMILY XV.—SPALACOPODIDÆ.

(*The Porcupines.*)

Body more or less armed with spines. Molars rooted, $\frac{4}{4}\cdot\frac{4}{4}$. Toes 4–5 in ours, sub-equal, with long, compressed, curved claws; soles warty. Muzzle hairy; upper lip without a groove; chiefly arboreal; nearly all are South American. The above characters apply rather to the sub-family *Cercolabinæ,* to which our genus belongs, than to the whole family.

1. ERETHIZON, F. Cuvier. American Porcupines.

1. ***E. dorsatus,*** (L.) F. Cuvier. White-haired Porcupine. Dark brown, spines chiefly white. N. Am., S. to Mexico.

FAMILY XVI.—LEPORIDÆ.

(*The Hares.*)

Incisors $\frac{4}{2}$, the extra pair in upper jaw small, and placed behind the principal pair, which are grooved in front; molars $\frac{6}{5}\cdot\frac{6}{5}$. A single well-known genus, widely distributed.

1. LEPUS, Linnæus. Hares.

* Fur white in winter.

1. ***L. americanus,*** Erxleben. White Rabbit. Northern Hare. Size large; hind feet longer than head; ears about equal to length of head; fur, in summer, cinnamon brown, in winter, becoming white at the surface, plumbeous at base, with a broad median band of reddish brown. Wooded districts, New England to Minn., and S. to Va., along the Alleganies.

** Fur never white.

2. ***L. sylvaticus,*** Bachman. Gray Rabbit. Size small; hind feet not longer than head; ears two-thirds length of head; gray above, varied with black, and more or less tinged with yellowish brown; below white. U. S. eastward; less northerly than the preceding. Two Southern species, *L. palustris*, Bach., the Marsh Rabbit, and *L. aquaticus*, Bach., the Water Rabbit, abound in S. Ills. (*Nelson.*)

ORDER F.—MARSUPIALIA.

(*The Marsupials.*)

FAMILY XVII.—DIDELPHIDIDÆ.

(*The Opossums.*)

Marsupial mammals of small size, with the teeth i. $\frac{5\cdot5}{4\cdot4}$, c. $\frac{1\cdot1}{1\cdot1}$, pm. $\frac{3\cdot3}{3\cdot3}$, m. $\frac{4\cdot4}{4\cdot4}$. Feet five-toed, plantigrade, claws 5–4. Tail usually very long, nearly naked, covered by a scaly skin, with a few scattered hairs, prehensile. All the species are American.

1. ***DIDELPHYS,*** Linnæus. OPOSSUMS.

1. ***D. virginiana,*** Shaw. COMMON OPOSSUM. Dirty white; legs dark; L. 35; T. 14. N. Y. to Rocky Mountains, rather southerly; common.

Class II.—Aves.

(*The Birds.*)

A Bird may be defined as an air-breathing vertebrate with a covering of feathers; warm blood; a complete double circulation; the two anterior limbs (wings) adapted for flying or swimming, the two posterior limbs (legs) adapted for walking or swimming; respiration never effected by gills or branchiæ, but, after leaving the egg, by lungs which are connected with air cavities in various parts of the body. Reproduction by eggs, which are fertilized within the body and hatched externally, either by incubation or exposure to the heat of the sun; the shell calcareous, hard and brittle.

Much more might be added, but the obvious distinction is this:—*All Birds have feathers, and no other animal has feathers.*

The classification of this group, as of most others, is still in an unsettled condition. Strictly speaking, the existing members of the class are so closely related that they might, with propriety, be combined into one order, which, by Prof. Gill, has been named Eurhipidura. At present, however, the term "order" may be applied to the groups so designated below, without thereby implying any such structural differences as exist between the "orders" of Reptiles or Fishes.

We now proceed to an artificial key to the

ORDERS OF BIRDS.

* Toes 3: two in front, one behind. . . . Picariæ, H.

** Toes 3: all in front; toes cleft or semipalmate. Limicolæ, M.

*** Toes 4: two in front, two behind.

Bill cered and hooked. . . . . Psittaci, I.

Bill lengthened, not cered nor hooked. Picariæ, H.

**** Toes 4: three in front, one behind.

I. Toes not webbed at all, cleft to the base, or with the basal joints immovably coherent.

a. Hind toe inserted on a level with the rest and generally longer than the shortest anterior toe.

b. Nostrils opening beneath a soft, swollen membrane; head small; tarsus reticulate behind. . Columbæ, K.

bb. Bill hooked and cered; claws sharp and strong. Raptores, J.

bbb. Secondaries very short, six in number; bill very slender; smallest of all birds. . . . Picariæ, H.

bbbb. Claw of hind toe as long or longer than that of middle toe; wing coverts in about two series, not reaching half way to tips of secondaries; musical apparatus more or less highly developed. . Passeres, G.

aa. Hind toe elevated above the level of the rest, and usually shorter than the others.

c. Bill fissirostral — culmen very short, but gape very wide and deep, reaching to below eyes. . Picariæ, H.

cc. Bill lengthened, not fissirostral.

d. First primary emarginate, or else about as long as second. . . . . . . Limicolæ, M.

dd. First primary not emarginate, much shorter than second. . . . . . . Alectorides, O.

II. Toes syndactyle — without webbing, but with the outer and middle toes coherent half their length. . Picariæ, H.

III. Toes semipalmate; two or three of them joined at base only by evident movable webbing.

e. Hind toe inserted on a level with the rest.

f. Tibiæ feathered below.

g. Bill cered and hooked; claws sharp and strong. Raptores, J.

gg. Bill not cered and hooked; nostrils opening beneath a soft, swollen membrane. . . Columbæ, K.

ff. Tibiæ naked below. . . . HERODIONES, N.

ee. Hind toe inserted above the level of the rest, and usually shorter than any of the others.

h. Tibiæ feathered below.

i. Nostrils perforate; head more or less naked. RAPTORES, J.

ii. Nostrils imperforate.

j. Bill fissirostral — gape wide, reaching to below eye. PICARIÆ, H.

jj. Bill stout, not fissirostral; nostrils scaled or feathered. GALLINÆ, L.

hh. Tibiæ naked below.

k. Nostrils perforate. . . . ALECTORIDES, O.

kk. Nostrils imperforate.

l. Head bald; tarsus reticulate. . HERODIONES, N.

ll. Head feathered; tarsus usually scutellate. LIMICOLÆ, M.

IV. Toes lobate, webbed at base or not, but *conspicuously* bordered on sides by plain or scalloped membranes.

m. Tail rudimentary; legs set far back. . PYGOPODES, S.

mm. Tail perfect; a horny frontal shield. ALECTORIDES, O.

mmm. Tail perfect; forehead feathered, without horny shield. LIMICOLÆ, M.

V. Toes palmate; three front toes full-webbed.

n. Bill curved upwards; legs elongated. . LIMICOLÆ, M.

nn. Bill lamellate, mostly flattish and furnished at tip with a decurved nail. . . . LAMELLIROSTRES, P.

nnn. Bill not recurved nor lamellate.

o. Hind toe not lobate; wings long and pointed; tail well developed. . . . . LONGIPENNES, R.

oo. Hind toe lobate; wings and tail short. PYGOPODES, S.

VI. Toes totipalmate; all four full-webbed. STEGANOPODES, Q.

G. FAMILIES OF PASSERES.

I. *Oscines.* Each side of tarsus covered with a plate, undivided in most of its length and meeting its fellow in a sharp ridge behind (in a few cases, back of tarsus without ridge, and formed of a few scutellæ distinct from those lapping over the front); first primary short, spurious or wanting, if present, not more than two-thirds of the longest; musical apparatus highly developed.

* Primaries 10; the first short or spurious.

† Tarsus booted; rictus with bristles.

a. Middle toe quite free from inner; birds of moderate size, length more than 6.

b. Wings moderate, not reaching when folded beyond the middle of tail, and not more than one-third longer than tail; tip of wing formed by 3d to 6th quill; no blue. . . . . TURDIDÆ, 18.

bb. Wings very long, pointed, reaching beyond middle of tail, and more than half longer; tip of wing formed by 2d to 4th quills; ours chiefly blue. SAXICOLIDÆ, 19.

aa. Middle and inner toes connected at base; small, length less than 5. . . . . SYLVIIDÆ, 20.

†† Tarsus scutellate in front.

c. Nostrils concealed by tufts of antrorse, bristly feathers.

d. First primary not more than half length of second; bill not notched; length less than 8.

e. Bill as long as head; wings pointed, much longer than tail. . . . . SITTIDÆ, 22.

ee. Bill much shorter than head; wings about as long as tail. . . . . PARIDÆ, 21.

dd. First primary more than half length of second; bill usually notched, the bristly nasal feathers branched to their tips; large, length more than 8. CORVIDÆ, 35.

cc. Nostrils exposed (rarely slightly overhung).

f. Bill distinctly notched near its tip, often hooked.

g. Tail longer than wings; general color gray or ashy-brown.

h. Bill very stout, compressed, strongly notched, toothed and abruptly hooked at tip; large, length 8 to 9. LANIIDÆ, 32.

hh. Bill more slender, not deeply notched nor abruptly hooked; length 8 to 10. . . TURDIDÆ, 18.

hhh. Bill very slender, not strongly notched nor hooked; small, length 4 to 5. . . SYLVIIDÆ, 20.

gg. Tail shorter than wings; general color olivaceous; bill stout, notched and hooked; length 4½ to 6½. VIREONIDÆ, 31.

ff. Bill not at all notched.

i. Rictus with bristles; quills not barred, the tail longer than wings; large, length 9 or more. TURDIDÆ, 18.

ii. No rictal bristles; wings and tail barred or undulated, usually about equal in length, the latter of rounded feathers; small, length 6 or less. TROGLODYTIDÆ, 24.

iii. No rictal bristles; tail about as long as wings, scansorial,—its feathers rigid and acute, not barred; bill long, decurved; length 5 to 6. . CERTHIIDÆ, 23.

** Primaries 9; the first about as long as second.

‡ Bill fissirostral,—triangular, depressed, about as wide at base as long; its wide, deep gape twice as long as the culmen, reaching to opposite the eyes; no rictal bristles; wings very long and pointed. . . . HIRUNDINIDÆ, 29.

‡‡ Bill tanagrine,—stout, conic, its outlines convex, the tomia with one or more lobes or nicks near the middle; nostrils very high; plumage brilliant, chiefly red (♂) or yellow (♀). TANAGRIDÆ, 28.

‡‡‡ Bill conirostral,—stout at base, and more or less conic; nostrils high up; tomia more or less evidently *angulated* near the base (*i. e.*, "corners of mouth drawn downward.")

j. Bill truly conic, much shorter than the head, usually notched at tip, or with bristles at the rictus. FRINGILLIDÆ, 33.

jj. Bill conic, but lengthened more or less, about as long as head (except in *Dolichonyx* and *Molothrus*, the Bob-o-link and Cowbird); no notch at the tip or bristles at the rictus. . . . . ICTERIDÆ, 34.

‡‡‡‡ Bill not as above, with the tomia straight, or very gently curved.

k. Conspicuously crested; bill triangular, depressed, notched, and hooked; tail tipped with yellow; secondaries (in full plumage) with red, horny tips. . AMPELIDÆ, 30.

kk. Nostrils concealed by bristly feathers; tarsus scutellate behind; hind claw long and nearly straight; inner secondaries lengthened. . . . ALAUDIDÆ, 25.

kkk. No crest; nostrils exposed; tarsus strictly "oscine."

l. Hind claw much elongated, twice as long as middle claw, with its toe much longer than middle toe and claw; bill very slender; longest secondary nearly equal to primaries in closed wing. . MOTACILLIDÆ, 26.

ll. Hind claw not specially elongated, not twice as long as middle claw; inner secondaries not lengthened.

m. Bill stout, compressed, notched, and abruptly hooked at tip; general color olivaceous, tail not blotched with white or yellow. . . VIREONIDÆ, 31.

mm. Bill various, notched or not, but little, if at all, hooked; colors often brilliant. SYLVICOLIDÆ, 27.

II. *Clamatores.* Outside of tarsus covered with a series of plates variously arranged, lapping entirely around in front and behind to meet in a groove on the inner side; primaries 10.

n. First primary lengthened, often longest, always more than $\frac{2}{3}$ length of the longest; bill broad, depressed, tapering to a point which is abruptly hooked; rictal bristles numerous; nostrils overhung but not concealed; tail not tipped with yellow. . . . . TYRANNIDÆ, 36.

H. FAMILIES OF PICARIÆ.

I. Feet zygodactyle (two toes in front, two behind) by reversion of outer toe; (hallux wanting in *Picoides.*)

a. Tail scansorial, of 12 rigid, acuminate feathers, of which the outer pair are short and concealed; bill stout and straight; nasal tufts usually developed. . . . PICIDÆ, 42.

aa. Tail not scansorial, of 8 to 10 long, soft feathers; bill decurved; no nasal tufts. . . . . CUCULIDÆ, 41.

II. Feet syndactyle, by cohesion of outer and middle toes; tibiæ naked below; bill stout and straight, longer than head. ALCEDINIDÆ, 40.

III. Feet neither zygodactyle nor syndactyle; wings long and pointed.

b. Bill tenuirostral, very slender, much longer than head; secondaries very short, 6 in number; plumage compact. TROCHILIDÆ, 39.

bb. Bill fissirostral, much shorter than head; secondaries more than 6.

c. Rictal bristles present; middle claw pectinate; plumage lax, variegated; length 8 or more. CAPRIMULGIDÆ, 37.

cc. No rictal bristles; plumage compact, of blended colors; tail feathers (in ours) spinous; length 6 or less. CYPSELIDÆ, 38.

I. FAMILIES OF PSITTACI.

I. Cere feathered, concealing the nostrils; plumage coarse and dry, chiefly green. . . . . ARIDÆ, 43.

J. FAMILIES OF RAPTORES.

I. Hind toe on a level with the rest, more than half length of outer toe, and with a large claw; claws strong, sharp, much curved; nostrils imperforate; head mostly feathered; bill strongly hooked.

a. Eyes directed forwards in consequence of the great lateral expansion of the cranium, and surrounded by a disk of radiating bristly feathers. . . . STRIGIDÆ, 44.

aa. Eyes lateral; no complete facial disk. . FALCONIDÆ, 45.

II. Hind toe elevated, not more than half length of outer toe; claws weak and little curved; nostrils perforate; head mostly naked; bill little hooked. . . . CATHARTIDÆ, 46.

K. FAMILIES OF COLUMBÆ.

I. Head small, feathered (except sometimes a circumorbital ring); feathers loosely inserted. . . . COLUMBIDÆ, 47.

L. FAMILIES OF GALLINÆ.

I. Head unfeathered, with wattles and caruncles; a tuft of bristly feathers on breast; tarsus spurred in ♂; plumage iridescent; large, 36 or more. . . . . MELEAGRIDÆ, 48.

II. Head feathered; plumage not iridescent; size much smaller.

Tarsus partly or entirely feathered, as is also the nasal groove; sides of neck usually with bare skin or peculiar feathers. . . . . TETRAONIDÆ, 49.

Tarsus and nasal groove unfeathered; no peculiar feathers on neck. . . . . PERDICIDÆ, 50.

M. FAMILIES OF LIMICOLÆ.

I. Toes lobate; tarsus notably compressed; body depressed. PHALAROPODIDÆ, 54.

II. Toes not lobate; tarsus not specially compressed.

* Legs exceedingly long; tarsus as long as tail; bill much longer than head, slender, acute, and curved upwards; feet 4-toed and palmate, or 3-toed and semipalmate. RECURVIROSTRIDÆ, 53.

** Bill usually shorter than head, pigeon-like, the broad, soft base separated by a constriction from the hard tip; head sub-globose, on a short neck; tarsus reticulate; toes 3 (except in *Squatarola*). . . . CHARADRIIDÆ, 51.

*** Bill usually longer than head, mostly grooved, not constricted, softish to its tip; tarsus scutellate; toes 4 (except in *Calidris*). . . . . . . SCOLOPACIDÆ, 55.

**** Not as above; bill hard, either compressed and truncate, or acute; feet 4-toed and cleft, or 3-toed and semipalmate. HÆMATOPODIDÆ, 52.

N. FAMILIES OF HERODIONES.

I. Bill long, straight, acute; middle claw pectinate. ARDEIDÆ, 56.

II. Bill curved downwards, or else flat and spoon-shaped. TANTALIDÆ, 57.

O. FAMILIES OF ALECTORIDES.

I. Very large; length 36 or more, with excessively long neck and legs; toes shorter than tarsus; bill contracted at the middle. GRUIDÆ, 58.

II. Smaller, length 18 or less, with comparatively short neck and legs; toes as long as tarsus; bill not contracted. RALLIDÆ, 59.

P. FAMILIES OF LAMELLIROSTRES.

I. Neck and legs moderate; tibiæ feathered; bill not decurved. ANATIDÆ, 60.

Q. FAMILIES OF STEGANOPODES.

I. Bill longer than tail, many times longer than head, with the gular pouch enormous; wings long. . PELECANIDÆ, 61.

II. Bill about as long as head, shorter than tail, which is fan-shaped, of rigid feathers; wings short. PHALACROCORACIDÆ, 62.

R. FAMILIES OF LONGIPENNES.

I. Nostrils not tubular, perforate; bill with a continuous covering. LARIDÆ, 63.

S. FAMILIES OF PYGOPODES.

I. Feet palmate; tail developed; head closely feathered. COLYMBIDÆ, 64.

II. Feet lobate; tail undeveloped; head usually with naked loral strip and peculiar feathers. . . . . PODICIPIDÆ, 65.

ORDER G.—PASSERES.

(*Passerine Birds.*)

Toes always 4; feet fitted for perching; the hind toe always on a level with the rest, its claw at least as long as that of middle toe, and often much longer; joints of toes respectively 2, 3, 4, 5, from first to fourth; toes never versatile; wing coverts comparatively few, chiefly in two series. Tail feathers 12, primaries 9 or 10. Musical apparatus more or less developed. Sternum of a certain uniform pattern. Nature altricial.

This group comprises the great majority of all Birds, and they represent the "highest grade of development, and the most complex organization of the class; their high physical irritability is co-ordinate with the rapidity of their respiration and circulation; they consume the most oxygen and live the fastest of all birds." (*Coues.*)

FAMILY XVIII.—TURDIDÆ.

(*The Thrushes.*)

Primaries 10, the first short or spurious; bill generally rather long, not conical, usually with a slight notch near the tip; nostrils oval, not concealed, but nearly or quite reached by the bristly frontal feathers; rictus with bristles, which are well developed in most of our species; tarsus in typical species, "booted," *i. e.*, enveloped in a continuous plate, formed by the fusion of all the scutellæ except two or three of the lowest; in other species distinctly scutellate. Toes deeply cleft, the inner one free, the outer united to the middle one, not more than half the length of the first basal joint.

A large family of more than two hundred species, found in most parts of the world, and embracing quite a wide variety of forms. Nearly all of them are remarkable for their vocal powers. Their food consists of insects and soft fruits.

Our species fall into three strongly marked sub-families, of which the *Miminæ* have been often associated with the Wrens, and the *Myiadestinæ* with the Wax Wings.

I. The TURDINÆ, or Typical Thrushes, have the tarsus booted, the first primary spurious, and the wings longer than the tail. They build rather rude nests, sometimes plastered with mud, and they lay four to six greenish or bluish eggs, either plain or speckled. All sing well, and some of them most exquisitely. Our species are usually referred to the typical genus, *Turdus*, but we have here separated the Wood Thrushes, as a group of full generic value (*Hylocichla*), as suggested by Prof. Baird. (Hist. N. Am. Birds, page 4.)

II. The MIMINÆ, or Mocking Thrushes, have the tarsus scutellate (sometimes booted in *Galeoscoptes*), the first primary scarcely spurious; the rictal bristles better developed, and the tail relatively longer, in our species longer than the wings. These birds have a brilliant and varied song, but all of them are plainly clad. All are American.

III. MYIADESTINÆ, the Fly-Catching Thrushes, have been usually associated with the *Ampelidæ*, but their affinities are rather with the thrushes, as Prof. Baird has shown. All are American,—the single species within our limits is a rare straggler from the West.

* Tarsus booted; wings longer than tail. (TURDINÆ.)

† Breast spotted; length 8½, or less. . . . HYLOCICHLA, 1.

†† Breast unspotted; (in ours) reddish or banded with black; length 9½, or more. . . . . . . TURDUS, 2.

** Tarsus scutellate in front (scutella rarely obsolete); wings (in ours) shorter than tail. (MIMINÆ.)

‡ Bill about as long as head, sometimes much longer, straight or curved, not notched. . . HARPORHYNCHUS, 3.

‡‡ Bill much shorter than head, notched at tip.

a. Tarsus distinctly scutellate; ours ashy, with black and white. . . . . MIMUS, 4.

aa. Tarsus feebly scutellate; plumage lead-colored; crissum chestnut-red. . . . . . . GALEOSCOPTES, 5.

*** Tarsus booted; wings about equal to tail; bill short, much depressed, notched and hooked; color ashy. (MYIADESTINÆ.) MYIADESTES, 6.

1. HYLOCICHLA, Baird. WOOD THRUSHES.

< *Turdus*, Linn.

1. ***H. mustelina,*** (Gm.) Bd. WOOD THRUSH. Cinnamon brown, brightest on the head, shading into olive on the

rump; breast with large, very distinct dusky spots; L. 8; W. 4¼; T. 3. E. U. S., in woodland; our largest and handsomest thrush. An exquisite songster.

2. ***H. pallasi,*** (Cab.) Bd. HERMIT THRUSH. Olive brown above, becoming rufous on rump and tail; breast with numerous, rather distinct, dusky spots; a whitish orbital ring; L. 7; W. 3½; T. 2½. N. Am., migrating early.

3. ***H. swainsoni,*** (Cab.) Bd. OLIVE-BACKED THRUSH. SWAINSON'S THRUSH. Uniform olive above; breast and throat thickly marked with large, dusky olive spots; breast and sides of head strongly buffy-tinted; a conspicuous buffy orbital ring; L. 7¼; W. 4; T. 3. N. Am.

4. ***H. aliciæ,*** Baird. GRAY CHEEKED THRUSH. ALICE THRUSH. Similar to the preceding, of which it is probably a variety, but without ring about eye, or any buffy tint about head. E. N. Am., ranging more northerly.

5. ***H. fuscescens,*** (Steph.) Bd. TAWNY THRUSH. WILSON'S THRUSH. VEERY. Uniform tawny above; breast and throat washed with brownish or pinkish yellow, and marked with small indistinct brownish spots; L. 7½; W. 4¼; T. 3⅙. E. N. Am., frequent, a fine songster.

2. *TURDUS,* Linnæus. THRUSHES.

* Sexes similar; breast not spotted nor banded; throat streaked; bill notched. (*Planesticus*, Bon.)

1. ***T. migratorius,*** L. ROBIN. AMERICAN RED BREAST. Olive gray above; head and tail blackish; throat white, with black streaks; under parts chestnut brown; L. 9¾; W. 5½; T. 4½. N. Am., abundant.

** Sexes unlike; throat unstreaked; male with a black collar; bill not notched. (*Hesperocichla*, Bd.)

2. ***T. nævius,*** Gm. OREGON ROBIN. VARIED THRUSH. Slate color, orange brown below; L. 9¾; W. 5; T. 4. Pacific slope, accidental in Mass., N. J., and L. I.

3. *HARPORHYNCHUS,* Cabanis. MOCKING THRUSHES.

1. ***H. rufus,*** (L.) Cab. BROWN THRUSH. SANDY MOCKING BIRD. THRASHER. Cinnamon red above; lower parts thickly spotted; bill nearly straight, shorter and much less curved than in many other *Harporhynchi*, five species of which occur in the U. S. beyond the Rocky Mountains; L. 11; W. 4; T. 5¼. E. U. S., abundant. A brilliant songster.

4. *MIMUS,* Boie. MOCKING BIRDS.

1. ***M. polyglottus,*** (L.) Boie. MOCKING BIRD. Ashy brown above; wings blackish, with white wing bars; tail blackish, outer feathers white; L. 9½; W. 4½; T. 5. U. S., chiefly southerly; N. to Mass., Iowa, etc. A renowned songster.

5. *GALEOSCOPTES,* Cabanis. CAT BIRDS.

< *Mimus*, Boie.

1. ***G. carolinensis,*** (L.) Cab. CAT BIRD. Dark slate color; crown and tail black; crissum brownish chestnut; L. 8¾; W. 3¾; T. 4. U. S., every where.

6. *MYIADESTES,* Swainson. FLY-CATCHING THRUSHES.

1. ***M. townsendi,*** (Aud.) Cab. TOWNSEND'S SOLITAIRE. Ashy gray, paler below; wing bands buffy; tail blackish; whitish ring about eye; young with reddish spots; L. 8; W. 4½; T. 4½. Rocky Mountains and westward, straying E. to Chicago. (*Nelson.*) An exquisite songster.

FAMILY XIX.—SAXICOLIDÆ.

(*The Stone Chats.*)

Characters similar to those of the Thrushes, but the wings longer and very much pointed, reaching, when folded, beyond the middle of the short tail. Tarsus "booted;" first primary spurious. A family scarcely distinct from *Turdidæ*, of about twelve genera and one hundred species. They are chiefly Old World birds, but two genera occurring in America. Ours are rather small (less than seven), with oval nostrils and bristles about the rictus.

* Chiefly or partly blue; tarsus not longer than middle toe and claw; bill stout. . . . . SIALIA, 1.

1. SIALIA, Swainson. BLUE BIRDS.

1. **S. sialis,** (L.) Haldeman. COMMON BLUE BIRD. Bright blue above, throat and breast reddish brown, belly white; ♀ usually duller with a brownish tinge on back; young, as in others, spotted; L. 6¾; W. 4; T. 3. E. N. Am., abundant; breeds every where.

2. **S. mexicana,** Sw. WESTERN BLUE BIRD. Head, neck all around and upper parts generally, blue; back with more or less chestnut; breast and sides reddish brown, otherwise bluish below; size of last. Pacific Slope, E. to Iowa (accidental.)

3. **S. arctica,** Sw. ROCKY MOUNTAIN BLUE BIRD. Rich greenish blue; belly white; ♀ with pale drab, instead of blue, on breast, etc.; size of others, or smaller. Central Table lands chiefly, E. to Missouri R.

FAMILY XX.—SYLVIIDÆ.

(*The Sylvias.*)

Primaries 10, the first short but scarcely spurious. Bill slender, depressed at base, notched and decurved at tip. Rictal bristles conspicuous; nostrils oval, overhung by a few bristles or a feather. Tarsus booted or scutellate. Basal joint of middle toe attached its whole length externally, half way internally. A large family of nearly six hundred species of small birds, chiefly of the Old World, where they take the place filled in America by the *Sylvicolidæ*. To this family belongs the European nightingale. Our species fall into two sub-families, *Regulinæ* and *Polioptilinæ*, each represented by its typical genus.

* Tarsus booted; wings longer than tail. . . REGULUS, 1.
** Tarsus scutellate; wings not longer than tail. POLIOPTILA, 2.

1. REGULUS, Cuvier. KINGLETS.

1. **R. satrapa,** Licht. GOLDEN-CROWNED KINGLET. Olivaceous; crown with a yellow patch, bordered with black, orange red in the center in ♂; extreme forehead and line over eye, whitish; vague dusky blotch at base of secondaries; a tiny feather over each nostril; L. 4; W. 2¼; T. 1¾. N. Am.

2. **R. calendula,** (L.) Licht. RUBY-CROWNED KINGLET. Olivaceous; crown with a scarlet patch in both sexes, wanting the first year; no black about head; no nasal feather; L. 4¼; W. 2⅛; T. 1¾. N. Am.

2. POLIOPTILA, Sclater. GNAT CATCHERS.

1. **P. cœrulea,** (L.) Sclater. BLUE-GRAY GNAT CATCHER. Clear ashy blue, brightest on head; whitish below; ♂ with forehead and sides of crown black; outer tail feathers chiefly white; L. 4⅛; W. 2; T. 2¼. U. S., chiefly southerly; N. to Mass. and L. Mich. Noticeable for its sprightly ways and squeaky voice, "like a mouse with the toothache;" but really a fine singer.

FAMILY XXI.—PARIDÆ.

(The Titmice.)

Primaries 10, first short; wings rounded; not longer than the rounded tail. Bill much shorter than head, not notched nor decurved at the tip; loral feathers bristly, and nostrils concealed by dense tufts. Tarsus scutellate, longer than middle toe and claw. Toes much soldered at base, widened beneath into a sort of palm. Plumage lax, little variable.

Small birds, less than seven inches long, resembling the Jays in several respects, restless, noisy, and scarcely migratory. Species seventy-five or more, chiefly of the Northern hemisphere, and abounding in both continents.

* Conspicuously crested; chiefly lead gray, paler below. LOPHOPHANES, 1.

** Not crested; crown, chin and throat black or brown. PARUS, 2.

1. *LOPHOPHANES,* Kaup. TUFTED TITMICE.

1. ***L. bicolor,*** (L.) Bon. TUFTED TITMOUSE. Forehead alone black; whitish below; sides washed with reddish; L. 6¼; W. 3¼; T. 3¼. E. U. S., southerly; N. to L. I. and L. Mich.; abundant in woodland and remarkable for its loud, ringing notes. Three other species occur in the S. W.

2. *PARUS,* Linnæus. CHICKADEES.

1. ***P. atricapillus,*** L. TITMOUSE. BLACK-CAPPED CHICKADEE. Grayish ash; wings and tail plain with whitish edging; crown, nape, chin and throat black; no white superciliary line; L. 5; W. 2½; T. 2½. N. Am.; abundant.

Var. ***carolinensis,*** (Aud.) Coues. SOUTHERN CHICKADEE. Smaller; tail feathers not noticeably white-edged. E. U. S.; southerly.

2. ***P. hudsonicus,*** Forster. HUDSONIAN CHICKADEE. Olive brown; crown browner; some pale chestnut below; a white superciliary line; L. 5; W. $2\frac{1}{8}$; T. $2\frac{2}{3}$. British America; S. to Mass.

FAMILY XXII.—SITTIDÆ.

(*The Nuthatches.*)

Primaries 10, the first spurious. Wings long and pointed, much longer than the broad soft tail. Bill not notched, rather slender, straight, nearly as long as head. Loral feathers bristly; nostrils concealed by dense tufts. Tarsus scutellate, shorter than middle toe and claw. Tongue acute, barbed. Body depressed; plumage lax, but less so than that of the Titmice. Active, nimble little birds, running up and down trees, and hanging in every conceivable attitude, the head down as often as up. Species twenty-five or thirty, in most parts of the world.

1. SITTA, Linnæus. NUTHATCHES.

1. ***S. carolinensis,*** Gm. WHITE-BELLIED NUTHATCH. "SAP SUCKER." Ashy blue above, white below; crissum, etc., washed with rusty brown; crown and nape black, unstriped; middle tail feathers like the back, others black, blotched with white; ♀ with less or no black on the head; L. $5\frac{1}{2}$; W. $3\frac{1}{2}$; T. 2. U. S.; abundant every where.

2. ***S. canadensis,*** L. RED-BELLIED NUTHATCH. Ashy blue, brighter than the preceding, rusty brown below; crown glossy black (♂), or bluish (♀), bordered by white and black stripes; L. $4\frac{1}{2}$; W. $2\frac{2}{3}$; T. $1\frac{1}{2}$. U. S., and northward.

3. ***S. pusilla,*** Lath. BROWN-HEADED NUTHATCH. Ashy blue; crown clear brown, a whitish spot on nape;

pale rusty below. L. 4; W. $2\frac{1}{2}$; T. $1\frac{1}{2}$. South Atlantic States.

FAMILY XXIII.—CERTHIIDÆ.

(*Creepers.*)

Primaries 10, first less than half second. Bill slender, as long as head; without notch or bristles, decurved. Tarsus scutellate, shorter than middle toe. Claws all very long, curved and compressed. Wings about as long as tail; tail feathers pointed, with stiffened shafts, almost wood-pecker like, and used for support in the same way. A small family of a dozen species, widely distributed. Habits similar to those of the Nuthatches, but the voice different, being small and fine. (The above diagnosis does not strictly apply to some foreign birds usually placed in this family.)

1. CERTHIA, Linnæus. Brown Creepers.

1. ***C. familiaris,*** Linn. Brown Creeper. Plumage dark brown, much barred and streaked; rump clear tawny; L. $5\frac{1}{2}$; W. $2\frac{3}{4}$; T. $2\frac{3}{4}$. N. Am. and Europe. A curious little bird, recognizable at once by the scansorial tail.

FAMILY XXIV.—TROGLODYTIDÆ.

(*The Wrens.*)

Primaries 10, the first short but hardly spurious. Wings rounded, usually about as long as the graduated tail. Bill more or less slender, usually elongated, not notched in any of our species. Nostrils oval, unbristled, overhung by a scale-like membrane. No rictal bristles. Loral feathers bristly. Tarsus scutellate. Lateral toes nearly equal; middle toe usually united to half the basal joint of inner toe, and to the whole of the basal joint of the outer, or more. Quills barred in most of our species.

A large family of small birds, chiefly belonging to Tropical America. Genera about sixteen; species one hundred or more. "Our species are sprightly, fearless and impudent little creatures, apt to show bad temper when they fancy themselves aggrieved by cats or people, or any thing else that is big or unpleasant to them; they quarrel a good deal, and are particularly spiteful towards martins and swallows, whose homes they often invade and occupy. Their song is bright and hearty, and they are fond of their own music; when disturbed at it they make a great ado with noisy scolding. Part of them (*Cistothorus*) live in reedy swamps and marshes, where they hang astonishingly big globular nests, with a little hole on one side, on tufts of rushes, and lay six or eight dark colored eggs; the others nest any where." (*Dr. Coues.*) They are all plainly colored, being chiefly brown. All are insectivorous, and most of them migratory.

* Back nearly uniform in color, a conspicuous white superciliary line; outstretched feet falling far short of end of tail. THRYOTHORUS, 1.

** Back barred crosswise, sometimes obscurely so; no conspicuous superciliary line; bill shorter than head; hind claw shorter than toe.

† Tail about as long as wings. . . . TROGLODYTES, 2.

†† Tail much shorter than wings. . . . ANORTHURA, 3.

*** Back streaked lengthwise, at least on shoulders; hind claw as long as the toe; tail barred. . . CISTOTHORUS, 4.

1. THRYOTHORUS, Vieillot. MOCKING WRENS.

* Tail not longer than wings, its feathers reddish brown with fine black bars. (*Thryothorus.*)

1. ***T. ludovicianus,*** (Gm.) Bon. CAROLINA WREN. Clear reddish brown, brightest on rump; tawny below;

L. 6; W. 2½; T. 2⅓. E. U. S., southerly; N. to Penn.; not migratory. A remarkable singer.

** Tail longer than wings, its feathers mostly black. (*Thryomanes.*)

2. ***T. bewickii,*** (Aud.) Bon. Bewick's Wren. Grayish brown; two middle tail feathers barred; L. 5½; W. 2¼; T. 2½. U. S., southerly; N. to Penn.

2. *TROGLODYTES,* Vieillot. Wrens.

1. ***T. aedon,*** Vieill. House Wren. Brown, brightest behind; rusty below; every where more or less waved with darker, distinctly so on wings, tail, etc.; L. 5; W. 2; T. 2. E. U. S.; abundant every where; very variable. < *T. domesticus,* (Bart.) Coues.

3. *ANORTHURA,* Rennie. Winter Wrens.

1. ***A. troglodytes,*** (L.) Coues. Winter Wren. Deep brown, waved with dusky; belly, wings and tail strongly barred; L. 4; W. 1⅞; T. 1¼. N. Am., northerly; U. S., in winter, not common. (*T. hyemalis,* Vieill.)

4. *CISTOTHORUS,* Cabanis. Marsh Wrens.

* Bill about half as long as head; no white superciliary line. (*Cistothorus.*)

1. ***C. stellaris,*** (Licht.) Cab. Short-Billed Marsh Wren. Dark brown, head and back darker; entire upper parts with white streaks; L. 4½; W. 1¾; T. 1¾. E. U. S., in marshes; rather rare.

** Bill slender, about as long as head; a conspicuous white superciliary line. (*Telmatodytes,* Cab.)

2. ***C. palustris,*** (Wilson) Baird. Long-Billed Marsh Wren. Clear brown; back with a black patch containing white streaks; otherwise unstreaked above; crown blackish; rump brown; L. 5; W. 2; T. 1¾. U. S.; abundant in reedy swamps.

FAMILY XXV.—ALAUDIDÆ.

(*The Larks.*)

First primary very short or entirely wanting. Tarsus scutellate in front and behind (a character singular among *Oscines.*) Bill short, of various forms in different species; nostrils concealed by tufts of antrorse feathers. Hind claw very long and nearly straight. Inner secondaries lengthened and flowing. A group of about one hundred species, chiefly Old World birds, but a single genus belonging to America; many of them are renowned as vocalists.

* Primaries 9; a little tuft of lengthened black feathers over each ear (sometimes obscure in ♀.) . . . EREMOPHILA, 1.

1. ***EREMOPHILA,*** Boie. HORNED LARKS.

= *Otocorys*, Bonap.

1. ***E. alpestris,*** (Forst.) Boie. SHORE LARK. Pinkish brown, thickly streaked; a crescent on breast and strip under eye black; white below; chin, throat, and line over eye more or less yellow; ♀ with less black; winter birds grayish, with the markings more obscure; L. 7¼; W. 4½; T. 3. N. Am. and Europe; common. A pleasant singer. [*E. cornuta*, (Wilson) Boie.]

FAMILY XXVI.—MOTACILLIDÆ.

(*The Wagtails.*)

Primaries 9, first about as long as second; inner secondaries enlarged, the longest one about as long as the primaries in the closed wing. Bill shorter than the head, very slender, straight, acute, notched at tip. Feet large, fitted for walking; hind claw long and nearly straight, inner toe cleft; basal joint of outer toe united with middle one. Rictal bristles not conspicuous; nostrils exposed.

A group of about one hundred species, mostly of the Old World, connecting the *Alaudidæ* with the *Sylvicolidæ*. Most of them are terrestrial. They have a habit (shared by various others) of moving the tail up and down, as if "balancing themselves on unsteady footing;" hence the name "Wagtail."

* Tarsus longer than middle toe and claw; outstretched feet falling much short of end of tail. . . . ANTHUS, 1.

1. ANTHUS, Bechstein. TITLARKS.

1. ***A. ludovicianus,*** (Gm.) Licht. BROWN LARK. TITLARK. PIPIT. Dark brown, slightly streaked; superciliary line and under parts buffy; breast and sides streaked; outer tail feathers more or less white; L. 6½; W. 3½; T. 3. N. Am.; generally abundant. (The Missouri Skylark, *Neocorys spraguei*, is a near relative.)

FAMILY XXVII.—SYLVICOLIDÆ

(The Warblers.)

Primaries 9; inner secondaries not enlarged, nor the hind toe long and straight, as in *Alaudidæ* and *Motacillidæ*. Bill usually rather slender, notched or not; the commissure not angulated at base, as in *Fringillidæ*, nor toothed in the middle, as in our *Tanagridæ;* the end not notched and abruptly hooked, as in *Vireonidæ* and *Laniidæ;* the gape not broad and reaching to the eyes, as in *Hirundinidæ*.

Our warblers are small birds; all (except *Icteria* which may not belong here) are less than six and a half inches in length, and very many are less than five. The rictus is generally bristled, but in several of our genera it is not. The colors are usually brilliant and variegated, but the sexes are unlike, and the variations due to age and season are great, so that the study of the species is often very difficult. Many of the Warblers are pleasing songsters, but none exhibit any remarkable powers in that line. All are insectivorous and migratory.

This family consists of more than a hundred species, chiefly North American, and embraces quite a wide variety, so that the group can perhaps be only distinguished negatively. The *Sylvicolidæ* grade perfectly into the *Tanagridæ* and *Cœrebidæ*, and probably the three families, and perhaps the *Fringillidæ*, also, should be merged into one. Our species are divisible into three very distinct sub-families, indicated below.

I. Bill slender, not hooked, as high as wide at base, with short bristles not reaching much beyond nostrils, or none; wings longer than tail (except *Geothlypis*); length 6½ or less. True Warblers. (SYLVICOLINÆ.)

* Tail feathers, some or all of them blotched with white.

† Rictus with evident bristles.

‡ Tarsus shorter than middle toe and claw; entirely black and white, streaked. . . . MNIOTILTA, 1.

‡‡ Tarsus not shorter than middle toe and claw.

a. Hind toe decidedly longer than its claw; bill acute, scarcely notched; bluish, throat and middle of back with yellow. . . . . . . PARULA, 2.

aa. Bill very acute, notched, perceptibly decurved, so that the gonys is slightly concave; rump and under parts chiefly yellow. . . PERISSOGLOSSA, 6.

aaa. Warblers without above characters. DENDRŒCA, 7.

†† Rictus without evident bristles.

b. Whole head and neck bright yellow; bill notched, half inch or more long. . . . PROTONOTARIA, 3.

bb. Whole head and neck not yellow; bill acute, not notched nor bristled, less than half inch long. HELMINTHOPHAGA, 5.

** Tail feathers yellow on inner webs; outer webs dusky; plumage chiefly yellow. . . . DENDRŒCA, 7.

*** Tail feathers all unmarked; same color on both webs.

c. Conspicuously streaked below; head plain or with two black stripes; legs long. . . . SEIURUS, 8.

cc. Not streaked below.

d. Wings about as long as tail; chiefly yellow below; crown (of ♂) black or ashy; legs strong. GEOTHLYPIS, 10.

dd. Wings decidedly longer than tail.

e. Bill not notched, half inch or more long; head plain or with four black stripes. . HELMITHERUS, 4.

ee. Bill less than half an inch long.

f. Bill notched; wings more than 2½; crown plain or with black. . . . . . OPORORNIS, 9.

ff. Bill not notched, nor bristled, very acute; wings less than 2½; crown plain or with a bright spot. HELMINTHOPHAGA, 5.

II. Bill rather stout, not notched, hooked nor bristled; tail longer than wings; length 7 to 8. Chats. (ICTERINÆ.) ICTERIA, 11.

III. Bill depressed, broader at base than high, notched and somewhat hooked, with strong rictal bristles half the length of bill; wings longer than tail; length 5½ or less. Fly-catching Warblers. (SETOPHAGINÆ.)

a. Bill fully twice as long as wide at base; tail feathers unmarked, or blotched with white. . MYIODIOCTES, 12.

aa. Bill scarcely twice as long as wide at base; tail marked with orange or yellow. . . . SETOPHAGA, 13.

1. MNIOTILTA, Vieillot. CREEPING WARBLERS.

1. ***M. varia,*** (L.) Vieill. BLACK AND WHITE CREEPER. Entirely black and white, streaked; crown with a broad white stripe; white wing bars; ♀ grayer; L. 5; W. 2¾; T. 2¼. E. U. S.; a neat bird, with some of the habits of a Nuthatch.

2. ***PARULA,*** Bonaparte. Blue Yellow-Backed Warblers.

= *Chloris*, Boie.
= *Sylvicola*, Sw. (Preoccupied in Mollusks.)

1. ***P. americana,*** (L.) Bon. Blue Yellow-Backed Warbler. Clear ashy blue; back with a large golden-green patch; yellow below, belly white; a brown band across breast; white wing bars. ♀ obscurely marked; L. 4¾; W. 2⅛; T. 2. Miss. Valley and E. One of our most elegant species, inhabiting tree-tops.

3. ***PROTONOTARIA,*** Baird. Golden Swamp Warblers.

1. ***P. citræa,*** (Bodd.) Bd. Prothonotary Warbler. Golden-Headed Warbler. Front and lower parts brilliant yellow; back, wings, etc., olivaceous; bill long; L. 5½; W. 3; T. 2¼. U. S., southward; N. to Wabash Valley, in bushy swamps, rather rare, one of the most beautiful of our birds.

4. ***HELMITHERUS,*** Rafinesque. Swamp Warblers.

1. ***H. vermivorus,*** (Gm.) Bon. Worm-Eating Swamp Warbler. Olive gree head yellowish, with four black stripes; buffy below; ♀ similar; L. 5½; W. 3; T. 2⅛. E. U. S.; N. to L. Erie.

5. ***HELMINTHOPHAGA,*** Cabanis. Worm-Eating Warblers.

* Tail feathers conspicuously blotched with white.

1. ***H. chrysoptera,*** (L.) Bd. Blue Golden-Winged Warbler. Ashy blue; forehead, crown and wing bars bright yellow; throat and broad stripe through eye, black, white below; ♀ duller; L. 5; W. 2½; T. 2¼. S. E. States, rather rare; N. to Green Bay; a beautiful species.

2. ***H. pinus,*** (L.) Bd. Blue-Winged Yellow Warb-

LER. Olive yellow; crown and all under parts bright yellow; wing bars whitish; loral strip black; ♀ similar; L. 4½; W. 2⅓; T. 2. S. E. States, N. to N. Y. A handsome bird, like a miniature *Protonotaria.*

** Tail feathers without white blotches.

3. ***H. ruficapilla,*** (Wils.) Bd. NASHVILLE WARBLER. Olive green, ashy on head and neck; crown patch bright chestnut, more or less concealed; bright yellow below; lores and orbital ring pale; ♀ duller, crown patch obscure; L. 4⅝; W. 2½; T. 2. E. U. S., frequent.

4. ***H. celata,*** (Say.) Bd. ORANGE-CROWNED WARBLER. Olive green, never ashy on head; crown patch orange brown, more or less concealed; greenish yellow below; ♀ duller, sometimes without crown patch; L. 4¾; W. 2¼; T. 2. Miss. Valley, S. & W.; rare E.

5. ***H. peregrina,*** (Wils.) Cab. TENNESSEE WARBLER. Olive green; no crown patch; white or slightly yellowish below; L. 4½; W. 2¾; T. 1¾. E. U. S., not common.

6. *PERISSOGLOSSA,* Baird. FRINGED TONGUE WARBLERS.

1. ***P. tigrina,*** (Gm.) Bd. CAPE MAY WARBLER. Olivaceous above with darker streaks; rump and sides of neck bright yellow; yellow below, much streaked with black; crown black or nearly so; ear coverts orange brown, a white wing patch; ♀ duller, with no black or reddish about head; L. 5¼; W. 2¾; T. 2. E. U. S., rather rare. A fine species with a peculiar structure of the tongue, which is somewhat as in *Cœrebidæ.*

7. *DENDRŒCA,* Gray. WOOD WARBLERS.

A large genus comprising about thirty species of brightly colored little birds, all American, and very abundant in the United States during the migrations.

Our species, though well marked, are often difficult to determine when not in full plumage. The tail feathers are always marked with white or yellow, and the bill is but moderately pointed, notched and with evident bristles at the rictus.

The following artificial analysis, partially borrowed from Coues' key to the genus, will generally enable the student to distinguish specimens.

* Tail feathers edged with yellow; plumage chiefly yellow. *æstiva*, 1.

** Tail feathers blotched with white.

† A white blotch on the primaries near their bases; no wing bars. . . . . *cærulescens*, 2.

†† No white blotch on primaries; wing bars, if present, not white.

White below; crown and wing patch more or less yellow. *pennsylvanica*, 6.

Yellow below; sides reddish-streaked; crown reddish. *palmarum*, 15.

Yellow below, sides black-streaked.

Back olive with reddish spots. . . . *discolor*, 12.

Back ashy. . . . . *kirtlandi*, 11.

††† No white blotch on primaries; wing bars or wing patch white.

‡ Rump yellow: —crown clear ash; yellow and streaked below. . . . . *maculosa*, 4.

—Crown with yellow spot; white and streaked below. . *coronata*, 3.

‡‡ Rump not yellow.

Crown with orange or yellow spot; throat orange or yellow. *blackburniæ*, 9.

Crown black; no distinct yellow any where; much streaked. *striata*, 8.

Crown blue or greenish, like the back; no definite yellow. *cærulea*, 5.

Crown chestnut, like the throat; no definite yellow; buffy below. . . . . *castanea*, 7.

Crown bluish or yellowish, not as above—some yellow.

Throat black (sometimes obscured by yellow tips to feathers); outer tail feather white-edged. *virens*, 13.

Throat yellow; —back ashy blue; cheeks black. *dominica*, 10.

—back yellowish olive; cheeks same. *pinus*, 14.

We copy from Coues' key the following valuable

DIAGNOSTIC MARKS OF WARBLERS IN ANY PLUMAGE.

A white spot at base of primaries. . . . *cærulescens*, 2.

Wings and tail dusky, edged with yellow. . . *æstiva*, 1.

Wing bars and belly yellow. . . . . *discolor*, 12.

Wing bars yellow and belly pure white. . *pennsylvanica*, 6.

Wing bars white and tail spots oblique, at end of two outer feathers only. . . . . *pinus*, 14.

Wing bars brownish; tail spots square at end of two outer feathers only. . . . . *palmarum*, 15.

Wing bars not evident (?); whole under parts yellow; back with no greenish. . . . . *kirtlandi*, 11.

Tail spots at end of nearly all the feathers, and no definite yellow any where. . . . . *cærulea*, 5.

Tail spots at middle of nearly all the feathers; rump and belly yellow. . . . . *maculosa*, 4.

Rump, sides of breast (usually) and crown with yellow; throat white. . . . . *coronata*, 3.

Throat definitely yellow; belly white; back with no greenish. *dominica*, 10.

Throat yellow or orange; crown with at least a trace of a central yellow or orange spot, and outer tail feather white-edged externally. . . . . *blackburniæ*, 9.

Throat, breast and sides black, or with black traces (seen on parting the feathers); sides of head with diffuse yellow; outer tail feather white-edged externally. . . . *virens*, 13.

With none of the foregoing special marks. *striata* 8 or *castanea* 7.

1. ***D. æstiva,*** (Gm.) Bd. SUMMER WARBLER. GOLDEN WARBLER. Chiefly golden yellow; back olive yellow; breast and sides with orange brown streaks; quills dusky, edged with yellow; ♀ similar, scarcely streaked; L. 5¼; W. 2½; T. 2¼. America; every where abundant.

2. ***D. cœrulescens,*** (L.) Bd. BLACK-THROATED BLUE WARBLER. Rich gray blue, with a few dusky streaks on back; throat, sides of head, neck and sides of body black, otherwise pure white below; quills black, edged with blue; ♀ dull olive greenish, obscurely marked, known by the blotch on the primaries; L. 5½; W. 2¾; T. 2¼. E. U. S.; an elegant species, not uncommon in woodland.

3. ***D. coronata,*** (L.) Gray. YELLOW-RUMPED WARBLER. MYRTLE WARBLER. Bluish ash above, streaked with black; white below with large blackish streaks; crown patch, rump and sides of breast bright yellow, there being four definite yellow places; ♀ and young brownish, with less yellow on breast and head; L. 5¾; W. 3; T. 2½. U. S., very abundant. The earliest migrant.

4. ***D. maculosa,*** (Gm.) Bd. BLACK AND YELLOW WARBLER. MAGNOLIA WARBLER. Back black, with olive skirtings; rump yellow; head clear ash; a white stripe behind eye; sides of head black, under parts (except the white crissum) rich yellow, with black streaks which are confluent on breast; ♀ similar, more olivaceous, with much less black; L. 5; W. 2½; T. 2¼. E. U. S. A brilliant little bird, common in woodlands.

5. ***D. cœrulea,*** (Wils.) Bd. CÆRULEAN WARBLER. Bright blue with black streaks; white below; breast and sides with bluish lines; ♀ not streaked, greenish above, slightly yellowish below; L. 4¼; W. 2½; T. 2. E. U. S.; N. to Niagara Falls; rather rare. A dainty species.

6. ***D. pennsylvanica,*** (L.) Bd. Chestnut-Sided Warbler. Blackish above; much streaked; crown clear yellow; black patch about eye; pure white below; a line of bright chestnut streaks along sides; wing patch yellowish (never clear white); ♀ similar but with less chestnut and black; L. 5; W. 2½; T. 2¼. E. U. S.; abundant, especially northward. A pretty species.

7. ***D. castanea,*** (Wils.) Bd. Bay-Breasted Warbler. Autumn Warbler. Back black and olive; thickly streaked; forehead and sides of head black enclosing a deep chestnut crown patch; chin, throat and sides, dull chestnut, otherwise pale buffy below; ♀ more olivaceous with less chestnut; young scarcely distinguishable from *striata;* L. 5; W. 3; T. 2½. E. U. S. Not very common.

8. ***D. striata,*** (Forst.) Bd. Black-Poll Warbler. Black and olivaceous, almost every where streaked; whole crown pure black; ♀ more olivaceous, slightly yellowish below; rather large; L. 5¾; W. 3; T. 2¼. E. U. S.; the last to migrate. "When the Black-Polls appear in force, the collecting season is about over." (*Coues.*)

9. ***D. blackburniæ,*** (Gm.) Bd. Orange-Throated Warbler. Blackburnian Warbler. Hemlock Warbler. Black above with whitish streaks; crown patch, superciliary line, sides of neck and the whole throat brilliant orange or flame color, fading into yellowish on the belly; ♀ similar, but olive and bright yellow instead of black and orange; L. 5¼; W. 2¾; T. 2¼. E. U. S.; abundant among the tree-tops. The most brilliant species.

10. ***D. dominica,*** (L.) Bd. Yellow-Throated Warbler. Ashy blue; throat bright yellow; belly white; cheeks black; superciliary line white or yellowish in front; L. 5; W. 2¾: T. 2⅓. Southern States; N. to

Penn., Central Indiana and Kansas; rare northward. A neat, plain species, with the habits of a creeper.

11. ***D. kirtlandi,*** Bd. Kirtland's Warbler. Ashy blue above; yellow and streaked below; lores black; L. 5½; W. 2¾; T. 2⅜. Ohio and the Bahamas. Two or three specimens known.

12. ***D. discolor,*** (Vieill.) Bd. Prairie Warbler. Olive yellow; back with a patch of red spots; forehead, superciliary line, wing bars and under parts bright yellow; streaked below; sides of head with black; ♀ similar; L. 4¾; W. 2¼; T. 2. E. U. S., N. to Mass. and Ills.; chiefly in evergreen thickets. An elegant species.

13. ***D. virens,*** (Gm.) Bd. Black-Throated Green Warbler. Clear yellow olive; sides of head rich yellow; whole throat and breast jet black, the color extending along the sides; otherwise whitish below; ♀ and winter birds with the black interrupted or veiled with yellowish; L. 5; W. 2½; T. 2¼. E. U. S.; abundant.

14. ***D. pinus,*** (Wils.) Bd. Pine-Creeping Warbler. Yellow olive above; under parts and superciliary line yellow; no sharp markings any where; ♀ more grayish; L. 5⅔; W. 3; T. 2⅓. E. U. S., N. to Mass. and L. Superior; abundant in evergreen forests.

15. ***D. palmarum,*** (Gm.) Bd. Yellow Red-Poll Warbler. Palm Warbler. Brownish olive above, somewhat streaked, rump brighter; crown bright chestnut; superciliary line and under parts yellow with brown streaks; no wing bars; ♀ similar; L. 5; W. 2⅔; T. 2¼. E. N. A.; abundant; terrestrial; less beautiful than most of the group.

8. SEIURUS, Swainson. Water Thrushes.

1. ***S. aurocapillus,*** (L.) Sw. Golden-Crowned Thrush. Oven-Bird. Bright olive green, white

below, sharply spotted on breast and sides, after the fashion of the Thrushes; crown orange brown, with two black stripes; L. 6¼; W. 3; T. 2½. U. S.; abundant in woodland, spending most of its time on the ground, like the other species of this genus, and the next two; remarkable for its loud, ringing song, and its curious oven-shaped nest; the largest of the true Warblers.

2. ***S. noveboracensis,*** (Gm.) Nutt. WATER WAGTAIL. WATER THRUSH. Dark olive brown above, pale yellowish beneath, thickly spotted every where with the color of the back; a yellowish superciliary line; bill about a half inch long; feet dark; L. 6; W. 3; T. 2⅛. N. Am.; abundant in low thickets; moves its tail like a Wagtail.

3. ***S. motacilla,*** (Vieill.) Bon. LARGE-BILLED WATER THRUSH. Same general color as last, but white or pale buffy below, and less sharply spotted; bill much larger, about ¾ inch; feet pale; larger; L. 6¼; W. 3¼; T. 2⅛. E. U. S., scarce; N. to Mass. (*Allen*) and N. Wis. (*Jordan.*) (*S. ludovicianus*, authors.)

9. *OPORORNIS,* Baird. NIMBLE WARBLERS.

1. ***O. agilis,*** (Wils.) Bd. CONNECTICUT WARBLER. Olive green, ashy on head; throat and breast brownish ash, otherwise yellow below; no sharp markings; in fall more olivaceous; L. 5¾; W. 3; T. 2¼. E. U. S., rare; a shy, quiet bird.

2. ***O. formosus,*** (Wils.) Bd. KENTUCKY WARBLER. Clear olive green, bright yellow below; crown and sides of head and neck black, with a rich yellow superciliary stripe, which bends around the eye behind; L. 5¾; W. 3; T. 2¼. E. U. S., chiefly southerly, N. to Wis. and Conn.; in low thickets, not generally common; a handsome and active species.

10. **GEOTHLYPIS,** Cabanis. GROUND WARBLERS.

1. ***G. trichas,*** (L.) Cab. MARYLAND YELLOW THROAT. BLACK-MASKED GROUND WARBLER. Olive green; forehead and broad mask extending down sides of head and neck jet black, bordered behind with clear ash; under parts yellow, clear on throat and breast; ♀ obscurely marked, without black mask and with less yellow; L. 4½; W. 2¼; T. 2⅛. U. S., abundant in thickets; a pretty bird with a lively song.

2. ***G. philadelphia,*** (Wils.) Bd. MOURNING WARBLER. Bright olive, clear yellow below; head ashy; throat and breast black, the feathers usually ashy-skirted (as though the bird wore crape, hence "Mourning Warbler"); ♀ and ♂ not in full plumage almost exactly like *O. agilis*, but the tail as long as wings; L. 5½; W. 2¼; T. 2¼. E. U. S., rather rare, in dense thickets.

11. **ICTERIA,** Vieillot. YELLOW-BREASTED CHATS.

1. ***I. virens,*** (L.) Bd. YELLOW-BREASTED CHAT. Olive-green; throat and breast bright yellow; belly abruptly white; lores black, a white superciliary line; wings and tail plain; tarsus almost booted; L. 7⅓; W. 3¼; T. 3⅓. U. S., southerly; N. to Mass. and Wis. (*Copeland.*) A loud, quaint songster, often placed with the *Vireos*, but having little affinity with any of our groups.

12. **MYIODIOCTES,** Audubon. FLY-CATCHING WARBLERS.

= *Wilsonia*, Nuttall (used in Botany).

1. ***M. mitratus,*** (Gm.) Aud. HOODED FLY-CATCHING WARBLER. YELLOW-MASKED WARBLER. Bright yellow-olive, crown and neck all around jet black, enclosing a broad golden mask; under parts from the breast bright yellow; tail with white blotches; ♀ olive instead of

black; L. 5; W. $2\frac{3}{4}$; T. $2\frac{1}{2}$. E. U. S., southerly; N. to L. Erie; a singular species.

2. ***M. pusillus,*** (Wils.) Bonap. GREEN BLACK-CAPPED FLY-CATCHING WARBLER. Clear yellow-olive; crown glossy black; forehead, lores, sides of head and entire under parts bright yellow; wings and tail unblotched; ♀ with less black; L. $4\frac{3}{4}$; W. $2\frac{1}{2}$; T. $2\frac{1}{4}$. U. S.; abundant.

3. ***M. canadensis,*** (L.) Aud. CANADA FLY-CATCHING WARBLER. Bluish ash; crown speckled with black; under parts (except white crissum) clear yellow; lores black, continuous with black under the eye, and this passing as a chain of black streaks down the side of the neck and encircling the breast like a necklace; wings and tail plain; ♀ similar, with less black; L. $5\frac{1}{3}$; W. $2\frac{3}{4}$; T. $2\frac{1}{2}$. E. U. S., to the Missouri, frequent. One of the handsomest Warblers.

13. SETOPHAGA, Swainson. AMERICAN REDSTARTS.

1. ***S. ruticilla,*** (L.) Sw. REDSTART. Black; sides of breast and large blotches on wings and tail orange-red; belly white, reddish tinged; ♀ olive, similarly marked with reddish yellow; L. $5\frac{1}{4}$; W. $2\frac{1}{2}$; T. $2\frac{1}{2}$. E. U. S., very abundant. A handsome and active Fly-Catcher.

FAMILY XXVIII.—TANAGRIDÆ.

(The Tanagers.)

Primaries 9; bill usually conical, sometimes depressed or attenuated, the culmen curved; cutting edges not much inflected, sometimes toothed, notched or serrated; tarsus scutellate. Legs short; claws long; colors usually brilliant. A large family of three hundred or more species, confined to the warmer parts of America, and embracing a wide diversity of forms. Some have slender bills and are scarcely distinguishable from the Warblers.

Others, like our *Pyranga*, have stout conical bills and are very closely related to the Finches. The single North American genus has a stout, sparrow-like bill, notched at the tip, and more or less evidently toothed or lobed near the middle of the upper mandible.

1. PYRANGA, Vieillot. FIRE TANAGERS.

1. ***P. rubra***, (L.) Vieill. SCARLET TANAGER. ♂ brilliant scarlet; wings and tail black, no wing bars; ♀ clear olive green; clear greenish yellow below; L. $7\frac{1}{3}$; W. 4; T. 3. E. U. S.; abundant in woodland; a respectable songster.

2. ***P. æstiva***, (L.) Vieill. SUMMER RED BIRD. ♂ bright rose red throughout; wings a little dusky; ♀ dull brownish olive, dull yellowish below; no wing bars; bill and feet paler than in *P. rubra;* size of last. E. U. S., chiefly southerly; N. to N. J. and Ills.; abundant.

FAMILY XXIX.—HIRUNDINIDÆ.

(*The Swallows.*)

Primaries 9; bill "fissirostral," *i.e.*, short, broad, triangular, depressed, the gape wide and about twice as long as the culmen, reaching to about opposite the eyes. Wings very long and pointed, the first primary usually longest, and twice as long as the last; secondaries very short. Tail more or less forked. Feet weak; tarsus scutellate, shorter than middle toe and claw. Plumage compact, and more or less lustrous.

A very natural family of about one hundred species, found all over the world. All are strong on the wing, insectivorous, and usually migratory.

* Plumage above more or less lustrous blue-black or green; no tarsal tuft nor recurved hooks on outer primary.

† Tail deeply forked; outer feathers attenuate and blotched with white. . . . . HIRUNDO, 1.

†† Tail scarcely forked; rump and forehead not colored like the back. . . . . PETROCHELIDON, 3.

††† Tail somewhat forked, unblotched; rump and forehead colored like the back.

a. Lustrous green or violaceous; pure white below; length less than 6½. . . . . . . TACHYCINETA, 2.

aa. Lustrous blue-black; ♀ paler and whitish below; length more than 6½. . . . . . PROGNE, 6.

** Plumage brownish-gray; scarcely lustrous and without shades of blue or green.

b. A little tuft of feathers on tarsus at base of hind toe; edge of wing smooth. . . . . COTYLE, 4.

bb. Outer web of first primary more or less saw-like, with a series of minute recurved hooks; no tarsal tuft. STELGIDOPTERYX, 5.

1. HIRUNDO, Linnæus. SWALLOWS.

1. ***H. horreorum,*** Barton. BARN SWALLOW. Lustrous steel-blue, pale chestnut below; forehead and throat deep chestnut; an imperfect steel-blue collar; tail very deeply forked; L. 7; W. 5; T. 4½. N. Am., abundant; breeding in colonies about barns, etc.

2. TACHYCINETA, Cabanis. WHITE-BELLIED SWALLOWS.

1. ***T. bicolor,*** (Vieill.) Coues. WHITE-BELLIED SWALLOW. Lustrous green, pure white below; ♀ duller; L. 6¼; W. 5; T. 2⅗. N. Am., abundant about water, nesting in trees, etc.; a handsome swallow.

3. PETROCHELIDON, Cabanis. CLIFF SWALLOWS.

1. ***P. lunifrons,*** (Say.) Cab. CLIFF SWALLOW. EAVE SWALLOW. Lustrous steel blue; forehead, sides of head, throat, rump, etc., of various shades of chestnut; a blue spot on breast, belly whitish; L. 5⅓; W. 4½; T. 2⅓. N. Am., abundant, formerly nesting in cliffs, but now under the eaves of barns, etc.

4. *COTYLE,* Boie. Bank Swallows.

1. ***C. riparia,*** (L.) Boie. Bank Swallow. Sand Martin. Dark gray, not iridescent, white below, a brown shade across the breast; L. $4\frac{3}{4}$; W. 4; T. 2. N. Am., abundant, breeding in holes in sandbanks, etc.

5. *STELGIDOPTERYX,* Baird. Rough-Winged Swallows.

1. ***S. serripennis,*** (Aud.) Bd. Rough-Winged Swallow. Brownish gray; wing hooks weak in ♀; L. $5\frac{1}{3}$; W. $4\frac{1}{2}$; T. $2\frac{1}{4}$. U. S., not common, rare eastward, breeding in banks, etc.

6. *PROGNE,* Boie. Martins.

1. ***P. subis,*** (L.) Bd. Purple Martin. Lustrous blue-black throughout; ♀ duller, whitish and streaky below; bill stout, almost hooked; L. $7\frac{1}{2}$; W. 6; T. $3\frac{1}{3}$. N. Am., abundant. (*P. purpurea*, Auct.)

FAMILY XXX.—AMPELIDÆ.

(*The Chatterers.*)

Primaries 10, or apparently 9, the first sometimes rudimentary and displaced; bill stout, triangular, depressed, decidedly notched and hooked, with the gape very wide. Nostrils overhung by membrane covered with bristly feathers. Tarsus short, with the lateral plates more or less subdivided, and often scarcely oscine in character; lateral toes nearly equal. As here constituted, a small group of six or eight species, the *Myidestinæ* usually brought into this connection being really *Turdidæ*, as shown by Prof. Baird. There are two sub-families, bearing but little resemblance to each other,—*Ptilogonydinæ*, of the warmer parts of N. America, and *Ampelinæ*, of the northern parts of both hemispheres.

The *Ampelinæ* constitute a single genus of three species. All are crested birds with a soft plumage of a handsome cinnamon drab color; the ends of the secondaries, and sometimes of the tail feathers, also, are tipped with horny appendages, looking like red sealing-wax.

The tail is short and square, much shorter than the long wings, and in our species it is tipped with yellow. The Wax Wings are migratory and gregarious, feeding on insects and soft fruits. Their voices are weak and wheezy, and they can scarcely be considered as songsters.

1. AMPELIS, Linnæus. WAX WINGS.

1. ***A. garrulus,*** L. BOHEMIAN WAX WING. NORTHERN WAX WING. General color an indescribable silky, ashy brown with a red tinge; front and sides of head shaded with purplish cinnamon; a black band across forehead around head; throat black; crissum chestnut red; two broad white wing bars; L. 7⅓; W. 4½; T. 3. Northern regions, S. in winter in large flocks to the Great Lakes; an interesting and beautiful bird.

2. ***A. cedrorum,*** (Vieill.) Bd. CEDAR BIRD. CHERRY BIRD. SOUTHERN WAX WING. Similar but smaller and less cinnamon-tinged, chin black; strip across face black, bordered above by whitish; belly yellowish; crissum white; no wing bars; ♀ with the wax-like appendages small or wanting; L. 6½; W. 3¾; T. 2½. E. U. S., abundant.

FAMILY XXXI. — VIREONIDÆ.

(*The Vireos.*)

Primaries 10, or apparently only 9, the first being sometimes rudimentary and displaced. Bill shorter than head, stout, compressed, decidedly notched and hooked.

Rictus with bristles. Nostrils exposed, overhung by a scale, reached by the bristly frontal feathers. Tarsus scutellate; toes soldered at base for the whole length of basal joint of middle one, which is united with the basal joint of the inner and the two basal joints of the outer; lateral toes usually unequal.

A rather small family, comprising sixty or seventy species of small olivaceous birds, all American. The coloration is usually blended and varies little with age or sex. All are insectivorous, and many of them are remarkable as songsters.

Concerning the "nine-primaried" species, Prof. Baird remarks: "In *V. flavifrons*, in which the outer primary is supposed to be wanting, its presence may be easily appreciated. One of the peculiar characters of this species consists in a narrow edging of white to all the primary quills, while the primary coverts (the small feathers covering their bases, as distinguished from what are usually termed the wing coverts, which more properly belong to the forearm or secondaries) are without them. If these coverts are carefully pushed aside, two small feathers considerably shorter than the others will be disclosed, one overlying the other, which (the under one) springs from the base of the exposed portion of the long outermost primary, and lies immediately against its outer edge. This small feather is stiff, falcate, and edged with white like the other quills, and can be brought partly around on the inner edge of the large primary, when it will look like any spurious quill. The overlying feather is soft, and without light edge.

In the other Vireos, with appreciable spurious or short outer primary, a similar examination will reveal only one small feather at the outer side of the base of the exterior large primary.

In all the families of Passeres, where the existence of nine primaries is supposed to be characteristic, I have invariably found, as far as my observations have extended, that there were two of the small feathers referred to, while in those of ten primaries but one would be detected."

* Wings long and pointed, ¼ or more longer than the tail; first primary very small or apparently wanting, less than ⅓ length of second. . . . . . . VIREOSYLVIA, 1.

** Wings relatively short and rounded, not one-fourth longer than the tail; first primary ⅖ or more length of second; bill stout. . . . . VIREO, 2.

1. *VIREOSYLVIA,* Bonaparte. LONG-WINGED VIREOS.

< *Vireo*, Vieillot.

* Slender species, the bill slender, light horn color, pale below; commissure straight and culmen relatively so; no wing bars nor conspicuous orbital ring; feet weak. (*Vireosylvia*)

† Primaries apparently 9.

1. ***V. olivacea,*** (L.) Bon. RED-EYED VIREO. GREENLET. Olive green, crown ashy, edged on each side with blackish; a white superciliary line, and below this a dusky streak; white below, somewhat olive shaded; eyes red; L. 6; W. 3⅓; T. 2½. E. U. S., very abundant in woodland; an energetic songster.

2. ***V. philadelphica,*** Cassin. PHILADELPHIA GREENLET. Dull olive green, becoming ashy on crown; no black lines on head; a whitish superciliary line; below faintly yellowish, fading to white on throat, etc.; L. 4¾; W. 2⅔; T. 2¼. E. U. S., scarce.

†† Primaries evidently 10.

3. ***V. gilva,*** (Vieill.) Cass. WARBLING VIREO. Colors exactly as in the preceding, but the spurious quill evident; L. 5⅓; W. 2¾; T. 2¼. E. N. A., frequent; an exquisite songster, nesting in tall trees in cities, etc.

** Stout species; the bill short and stout, blue-black; both culmen and commissure decidedly curved; a pale stripe running from bill to and around eye; white wing bars; quills blackish, mostly edged with white; feet stout. (*Lanivireo*, Bd.)

‡ Primaries apparently 9.

4. ***V. flavifrons,*** (Vieill.) Baird. Yellow-Throated Vireo. Rich olive green above, becoming ashy on rump; bright yellow below; belly white; superciliary line and orbital ring yellow; L. 5¾; W. 3; T. 2. E. U. S., abundant; a brightly colored species.

‡‡ Primaries evidently 10.

5. ***V. solitaria,*** (Wils.) Baird. Blue-Headed Vireo. Solitary Greenlet. Bright olive green; crown and sides of head bluish ash; stripe to and around eye white, a dusky line below it; white below, somewhat washed with pale yellow; L. 5⅝; W. 3; T. 2⅓. U. S., in woodland, frequent; a stout, handsome species.

2. *VIREO,* Vieillot. Short-Winged Vireos.

1. ***V. noveboracensis,*** (Gm.) Bonap. White-Eyed Vireo. Bright olive green, white below; sides and crissum bright yellow; pale wing bars; stripe from bill to and around eye, yellow; eyes white; L. 5; W. 2⅓; T. 2¼. E. U. S., in thickets; a sprightly bird, with a loud and varied song.

2. ***V. belli,*** Aud. Bell's Vireo. Olive-green, yellow below, chin and superciliary line whitish; wing bars whitish; L. 4¼; W. 2⅕; T. 2. Western, E. to Ills. and Neb. Resembles *V. gilva.*

FAMILY XXXII.—LANIIDÆ.

(*The Shrikes.*)

Primaries 10, the first short (rarely wanting); bill hawk-like, very strong, the upper mandible toothed and abruptly hooked at the tip; both mandibles distinctly notched. Wings short, rounded. Tail long. Tarsus scutellate on the outside as well as in front. Sexes alike.

Species about 100, found in most parts of the world, remarkable for their energy and pugnacity.

* Rictus with bristles; nostrils concealed by bristly tufts; colors black, white and gray. . . . . Collurio, 1.

1. COLLURIO, Vigors. Shrikes.

1. ***C. borealis***, (Vieill.) Baird. Great Northern Shrike. Butcherbird. Clear bluish ash above; black bars on side of head *not* meeting in front, interrupted by a white crescent on under eyelid; rump and shoulders whitish; wings black; white below, waved with blackish; L. $9\frac{1}{2}$; W. $4\frac{1}{4}$; T. $4\frac{3}{4}$. Northern regions, S in winter to Ohio R. and Potomac.

2. ***C. ludovicianus***, (L.) var. ***excubitoroides***, (Sw.) Coues. Loggerhead Shrike. Clear ashy blue; a whitish superciliary line; black bars on sides of head meeting across forehead; no crescent on under eyelid; white below scarcely or not dark-waved; L. $8\frac{1}{4}$; W. 4; T. $4\frac{1}{4}$. Western, E. to L. Michigan and Ohio R.

FAMILY XXXIII.—FRINGILLIDÆ.

(*The Finches.*)

Primaries 9. Tarsus strictly oscine. Bill mostly shorter than head, robust, of a conical form, with the commissure more or less abruptly angulated near its base; in other words, the "corners of the mouth drawn down." This feature is usually unmistakeable, and it is almost the only character pertaining to all the members of the family. Even this is also shared by the *Icteridæ*, which, however, may generally be distinguished by the greater length and slenderness of the bill.

A very large family, the most extensive in Ornithology, comprising about one hundred genera and five hundred species, found in nearly every part of the world, except Australia. They are especially abundant in North America, where about one-eighth of all the birds are *Fringillidæ.* "Any one United States locality of average attractiveness to birds, has a bird-fauna of over two hundred species, and if it be away from the sea-coast, and consequently uninhabited by marine birds, about one-fourth of the species are *Sylvicolidæ* and *Fringillidæ* together, the latter somewhat in excess of the former. It is not easy, therefore, to give undue prominence to these two families." (*Coues.*)

All the Finches are granivorous, feeding chiefly on seeds, but not rejecting either berries or insects; nearly all sing, and some most delightfully; most of them are plainly clad, a streaky brown being the prevailing tint, but others are among the most brilliantly colored birds. Among these latter only are the changes in plumage strongly marked.

The following key to the genera is about as artificial as it well could be, but a more natural one would be less easy of application. The characters here assigned are seldom truly generic.

* Species of large size; length at least more than 7½.

† Tail longer than wings.

a. Conspicuously crested, chiefly red or rosy-tinted; bill very large, reddish. . . . . . CARDINALIS, 23.

aa. Not crested; black or brown with chestnut on sides; wings and tail with white; bill moderate, black. PIPILO, 24.

aaa. Not crested, head mostly black; no white on tail. ZONOTRICHIA, 13.

†† Tail shorter than wings.

b. Bill very large and stout. ("*Grosbeaks.*")

c. Black and white (♂) or brown, streaked (♀); under wing coverts rosy or yellow. . . . GONIAPHEA, 20.

cc. Rosy red (♂) or gray with brownish yellow on head and rump (♀). . . . . PINICOLA, 2.

ccc. Bill greenish yellow, as long as tarsus; wings and tail black; secondaries mostly white. HESPERIPHONA, 1.

bb. Bill moderate or small.

d. White, with black on wings and tail, or washed with clear brown; hind toe elongated. PLECTROPHANES, 7.

dd. Streaked above; head striped; tail about as long as wings. . . . . . . ZONOTRICHIA, 13.

** Species of medium or small size; length 7¼ or less.

‡ Mandibles long and much curved, their points crossed; colors chiefly red or olive. . . . . LOXIA, 4.

‡‡ Hind claw straightish, twice as long as middle claw; colors black, white and brown. . . . PLECTROPHANES, 7.

‡‡‡ With neither of the preceding combinations.

e. No where *decidedly* spotted or streaked (sometimes appearing mottled owing to the darker centers of the feathers).

f. Blackish, or ashy; belly and one to three outer tail feathers white; bill pale, without ruff. . JUNCO, 17.

ff. Yellow, more or less; base of bill with a small ruff; no blue; young brownish. . . CHRYSOMITRIS, 6.

fff. Chiefly or entirely blue (♂), greenish or plain brown (♀).

g. Length more than 6; wings with chestnut or whitish; bill stout. . . . . . . GUIRACA, 21.

gg. Length 5 to 6; gonys usually with a dusky stripe. CYANOSPIZA, 22.

ee. Some where or every where decidedly spotted or streaked.

h. One or more outer tail feathers partly or wholly white.

i. Hind claw very long and nearly straight; colors black and white or brown. . . PLECTROPHANES, 7.

ii. Hind claw not specially elongated.

j. Bend of wing with chestnut; crown and breast streaked; tail much shorter than wings. POŒCETES, 10.

jj. No chestnut on wing; breast unstreaked; head with black, white and chestnut; tail nearly as long as wings. . . . . . . CHONDESTES, 12.

hh. Tail feathers rigid, acute, almost scansorial; small streaked marsh-sparrows with yellow-edged wings. AMMODROMUS, 11.

hhh. Tail feathers more or less rounded and soft, none of them white.

k. Wings *decidedly* longer than tail.

l. With crimson or clear (not rusty) red; a ruff at base of bill.

m. Crown crimson; throat dusky. . ÆGIOTHUS, 5.

mm. Crown, chin, throat and often whole plumage washed with red. . . ♂ of CARPODACUS, 3.

ll. With definite yellow some where.

n. Bases and edges of quills and tail feathers yellow; bill acute. . . . CHRYSOMITRIS, 6.

nn. Rump sulphur yellow; bill with a small ruff. ÆGIOTHUS, 5.

nnn. Edge of wing and superciliary line or spot at least, yellow or yellowish; no ruff.

o. Breast yellow; throat patch or streaks black, bill bluish. . . . . . EUSPIZA, 19.

oo. Breast buffy or streaky; wings less than 2½; tail feathers narrow. . AMMODROMUS, 11.

ooo. Breast streaked; wings more than 2½; inner secondaries nearly as long as primaries. PASSERCULUS, 9.

lll. With no definite crimson nor yellow any where.

p. Introduced birds, not streaked below; throat black in ♂. . . . . PASSER, 8.

pp. Native birds, much streaked below.

q. Inner claw reaching at least half way to tip of middle claw; tail, wings, etc., with much chestnut red; wings more than 3; no ruff. PASSERELLA, 18.

qq. Olivaceous; no black nor chestnut; wings more than 3; secondaries not lengthened; a ruff at base of bill. . . ♀ of CARPODACUS, 3.

qqq. Inner secondaries lengthened, about as long as primaries; wings less than 3; no ruff. PASSERCULUS, 9.

kk. Wings little if any longer than tail.

r. Tail feathers very slender, rather stiff and sharp pointed. . . . . AMMODROMUS, 11.

rr. Tail feathers not rigid and sharp pointed.

s. Sharply streaked below. . . MELOSPIZA, 15.

ss. Not streaked below (when adult.)

t. Crown chestnut in adult (streaky in young); no yellow.

u. Tail rounded; length about 5¾; wings and tail less than 2½; sharply streaked above. MELOSPIZA, 15.

uu. Tail forked; length 5 to 6¼; wings and tail 2½ to 3; tarsus ⅔ to ¾. . SPIZELLA, 14.

tt. Crown not chestnut in adult, often partially so in young.

v. Head striped; length more than 6; tarsus more than ¾. . . ZONOTRICHIA, 13.

vv. Length less than 6; bend of wing yellowish. PEUCÆA, 16.

1. HESPERIPHONA, Bonaparte. EVENING GROSBEAKS.

1. ***H. vespertina,*** (Coop.) Bon. EVENING GROSBEAK. Olivaceous; crown, wings, tail and tibia black; forehead and crissum yellow; bill very large, yellowish; L. 8; W. 4¼; T. 2½. Western, E. to Ohio, etc.

2. PINICOLA, Vieillot. PINE GROSBEAKS.

1. ***P. enucleator,*** (L.) Vieill. PINE GROSBEAK. ♂ chiefly red; white wing bars; ♀ ashy gray with brownish yellow on head and rump; L. 8½; W. 4½; T. 4. Northward, S. in winter; in pine woods, etc.

3. *CARPODACUS*, Kaup. Purple Finches.

1. ***C. purpureus***, (Gmel.) Gray. Purple Finch. Every where streaky; ♂ flushed with red, most intense on the crown, fading below and behind; ♀ olive brown with no red; bill stout; L. 6; W. 3⅓; T. 2½. U. S., a fine songster.

4. *LOXIA*, Linnæus. Crossbills.

1. ***L. leucoptera***, (Gmel.) White Winged Crossbill. ♂ rose red; white wing bars; ♀ brownish olive, speckled with dusky; rump yellow; L. 6¼; W. 3½; T. 2½. Northern, S. in winter.

2. ***L. curvirostra***, L. Red Crossbill. ♂ brick-red; wings unmarked; ♀ brownish olive; L. 6; W. 3⅓; T. 2½. Northern regions and pine woods; S. in winter.

5. *ÆGIOTHUS*, Cabanis. Linnets

1. ***Æ. linarius***, (L.) Cab. Red Poll Linnet. Crown crimson in both sexes; throat, breast and rump also rosy in ♂; much streaked above; chin blackish; L. 5¾; W. 3; T. 2½. Northern, S. in winter, in flocks.

2. ***Æ. flavirostris***, (L.) var. ***brewsteri***, Ridgway. Brewster's Linnet. No red on crown or breast; rump rosy in ♂; yellow in ♀; L. 5½; W. 3; T. 2½. Mass., lately discovered.

6. *CHRYSOMITRIS*, Boie. Goldfinches.

* Sexes alike; plumage thickly streaked every where; no black on head; bill very sharp. (*Chrysomitris.*)

1. ***C. pinus***, (Wils.) Bon. Pine Linnet. Plumage streaky brown, suffused with yellow in the breeding season; bases of quills and tail feathers yellow, much as in the female Redstart; L. 4¾; W. 2¾; T. 2. N. Am.. rather northward. but liable to "turn up" any where.

** Sexes unlike; scarcely or not streaked; adult ♂ with black on crown, wings and tail. (*Astragalinus*, Cab.)

2. ***C. tristis,*** (L.) Bon. YELLOW BIRD. THISTLE BIRD. AM. GOLDFINCH. ♂ rich yellow; rump whitish; wing bars white; white spot on each tail feather; ♀ more olivaceous; fall plumage pale yellow brown; young variously ochraceous, with yellow or not; L. 5; W. 3; T. 2. N. Am.; every where.

7. *PLECTROPHANES,* Meyer. LONGSPURS.

* Bill small, with a ruff; hind claw long but curved. (*Plectrophanes.*)

1. ***P. nivalis,*** (L.) Meyer. SNOW BUNTING. SNOW FLAKE. In breeding season, pure white, with black on back, wings and tail; bill and feet black; in U. S. usually bill pale, and white of body clouded with clear, warm brown; L. 7; W. 4½; T. 3. Northern, S. in winter to Ohio R.; a beautiful bird.

** Bill larger, without ruff; hind claw nearly straight. (*Centrophanes*, Kaup.)

2. ***P. lapponicus,*** (L.) Selby. LAPLAND LONGSPUR. ♂ with head and throat mostly black; a chestnut collar; back black and streaky, whitish below; outer tail feathers with white; legs and feet black; ♀ and winter birds with less black; L. 6¼; W. 4; T. 2¾. Northern, S. in winter to N. Y. and Ills.

3. ***P. pictus,*** Sw. PAINTED LARK BUNTING. ♂ with head and upper parts mostly black; collar and under parts rich fawn color; legs pale; ♀ duller. Northern, S. in the interior to Ills. and Kans.; rare.

8. *PASSER,* Brisson. HOUSE SPARROWS.
= *Pyrgita*, Cuvier.

1. ***P. domesticus,*** L. ENGLISH SPARROW. ♂ chestnut

brown above, thickly streaked; ashy below; throat, lores and chin black; ♀ duller, without black; feet small; L. 6; W. 2¾; T. 2½. Introduced from Europe; abundant in the large cities.

2. ***P. montanus,*** Auct. EUROPEAN TREE SPARROW. "Distinguished by the chestnut crown, and the similarity of both sexes and the young." Introduced with preceding, and abundant in St. Louis (*Dr. J. C. Merrill*), and perhaps other places.

9. PASSERCULUS, Bonaparte. SAVANNA SPARROWS.

1. ***P. savanna,*** (Wils.) Bon. SAVANNA SPARROW. Sharply streaked; streaks on back blackish; superciliary line and edge of wing yellowish; L. 5½; W. 2¾; T. 2. N. Am., abundant on plains and shores.

2. ***P. princeps,*** Maynard. IPSWICH SPARROW. Streaks on back sandy brown, not sharply defined; superciliary line white in front; L. 6; W. 3¼; T. 2½. Mass., lately discovered.

10. POŒCETES, Baird. GRASS SPARROWS.

1. ***P. gramineus,*** (Gm.) Baird. BAY-WINGED BUNTING. GRASS SPARROW. GROUND BIRD. Thickly streaked every where; slightly buffy below; L. 6; W. 3; T. 2½. N. Am., abundant in fields, etc., and known at once by the chestnut bend of wing and white outer tail feathers.

11. AMMODROMUS, Swainson. SHORE SPARROWS.

* Bill stout; tail feathers acute but not rigid; crown with a medium light stripe; inland species. (*Coturniculus*, Bon.)

1. ***A. passerinus,*** (Wils.) Baird. YELLOW-WINGED SPARROW. Much streaked above; feathers edged with bay; breast buffy, unstreaked; wings and tail short; edge

and bend of wing and line over eye yellow; L. 5; W. $2\frac{2}{3}$; T. 2. U. S., in fields; notes sharp, grasshopper-like.

2. ***A. henslowi,*** (Aud.) Baird. HENSLOW'S SPARROW. Smaller; more yellow above; breast, etc., with some sharp black streaks; L. 5; W. $2\frac{1}{4}$; T. $2\frac{1}{5}$. E. U. S., scarce; N. to Mass.

3. ***A. lecontei,*** (Aud.) Baird. LE CONTE'S SPARROW. Intermediate between the preceding and the next; bill small, blue-black; back with rufous; tail feathers very sharp and slender; breast unspotted, a broad buffy superciliary stripe; L. $4\frac{2}{3}$; W. $2\frac{1}{5}$; T. $2\frac{1}{3}$. Chicago, Ills. (*E. W. Nelson*) to Texas and N. W.; very rare.

** Bill long and slender; tail feathers sharp and rather stiff; seashore Sparrows. (*Ammodromus.*)

4. ***A. maritimus,*** (Wils.) Sw. SEA-SIDE FINCH. Olive gray; back obscurely streaked; a yellow spot over eye; L. 6; W. $2\frac{1}{3}$; T. 2. Salt marshes, Atlantic coast.

5. ***A. caudacutus,*** (Gm.) Sw. SHARP-TAILED FINCH. Back sharply streaked; no yellow spot over eye, a bright buff superciliary stripe; L. 5; W. $2\frac{1}{3}$; T. $1\frac{3}{4}$. Atlantic coast.

Var. ***nelsoni,*** Allen. NELSON'S SHARP-TAILED FINCH. Shores of Lake Michigan about Chicago; common. (*E. W. Nelson.*)

12. ***CHONDESTES,*** Swainson. LARK SPARROWS.

1. ***C. grammaca,*** (Say.) Bon. LARK FINCH. Streaked above, ashy below; crown and ear coverts chestnut, blackening on forehead, with whitish median and superciliary stripes; black lines through and below eye; and a conspicuous black line on each side of the white throat; a black pectoral spot; middle tail feathers like back, the rest blackish, white tipped; L. $6\frac{1}{2}$; W. $3\frac{1}{2}$; T.

3. Western, E. to Ohio; abundant on prairies and river bluffs; a fine songster.

13. *ZONOTRICHIA,* Swainson. White-Crowned Sparrows.

1. ***Z. leucophrys,*** (Forst.) Sw. White-Crowned Sparrow. Streaked above, with but little chestnut; crown with a broad white median band, a narrow black one and a white one on each side of it; no yellow any where; throat like breast; young with the crown chiefly rich brown; L. 7; W. 3¼; T. 3¼. N. Am.; less common than the next.

2. ***Z. albicollis,*** (Gm.) Bon. White-Throated Sparrow. Peabody Bird. Much chestnut streaking above; crown black, with white median and superciliary stripes; spot over eye and edge of wing always yellow; ashy below, whitening on throat; ♀ duller; L. 7; W. 3; T. 3⅙. E. N. Am.; an abundant and handsome sparrow.

3. ***Z. querula,*** (Nutt.) Gambel. Black-Hooded Sparrow. Crown, face and throat jet black; no yellow; ♀ with less black; L. 7½; W. 3¼; T. 3½. Missouri region, E. to Minn.

14. *SPIZELLA,* Bonaparte. Chipping Sparrows.

1. ***S. monticola,*** (Gm.) Baird. Tree Sparrow. Streaked above; crown chestnut; bill black above, yellow below; neck, line over eye and under parts ashy gray; a dark pectoral blotch; white wing bars; L. 6¼; W. 3; T. 3. N. Am., chiefly northerly; U. S. in winter.

2. ***S. pusilla,*** (Wils.) Bon. Field Sparrow. General color of preceding, but paler and duller; bill pale; wing bands rather obscure; L. 5½; W. 2⅓; T. 2⅓. E. U. S., abundant. [*S. agrestis,* (Bart.) Coues.]

3. ***S. socialis,*** (Wils.) Bon. CHIPPY. HAIR BIRD. Streaked above, with much dull bay; crown chestnut; bill, forehead and streak through eye black; ashy below; L. $5\frac{1}{4}$; W. $2\frac{2}{3}$; T. $2\frac{1}{2}$. N. Am., every where. [*S. domestica*, (Bart.) Coues.]

4. ***S. pallida,*** (Sw.) Bon. CLAY-COLORED SPARROW. Still smaller; pale brownish yellow, streaked with black; crown grayish, with median stripe. S. Ills. and W.

15. MELOSPIZA, Baird. SONG SPARROWS.

1. ***M. melodia,*** (Wils.) Baird. SONG SPARROW. Much streaked above, and on breast and sides; crown with an obscure pale median stripe; below white, pectoral streaks often forming a blotch; L. $6\frac{1}{2}$; W. $2\frac{1}{2}$; T. 3. U. S., every where; a well-known songster. [*M. fasciata*, (Gmel.) Scott.]

2. ***M. palustris,*** (Wils.) Baird. SWAMP SPARROW. Crown chestnut; wings strongly tinged with chestnut; breast and below with few streaks or none; tail shorter than in the Song Sparrow; L. $5\frac{3}{4}$; W. $2\frac{1}{3}$; T. $2\frac{1}{3}$. E. U. S., in low thickets.

3. ***M. lincolni,*** (Aud.) Baird. LINCOLN'S FINCH. Every where thickly, narrowly and sharply streaked; breast with a broad band of pale buffy or yellowish brown; sides washed with the same; L. $5\frac{1}{2}$; W. $2\frac{1}{2}$; T. $2\frac{1}{2}$. N. Am., rare eastward; a shy species quite unlike the others.

16. PEUCÆA, Audubon. SUMMER SPARROWS.

1. ***P. æstivalis,*** (Licht.) Cab. BACHMAN'S FINCH. Much streaked above, ashy below; yellow on bend of wing but none on head; L. 6; W. $2\frac{1}{3}$; T. $2\frac{1}{2}$. Southern, N. to Illinois.

17. ***JUNCO,*** Wagler. Snow Birds.

1. ***J. hyemalis,*** (L.) Scl. Snow Bird. ♀ more grayish; L. 6¼; W. 3; T. 3. E. N. Am., every where abundant, mostly seen in winter.

18. ***PASSERELLA,*** Swainson. Fox Sparrows.

1. ***P. iliaca,*** (Merrem.) Sw. Fox Sparrow. Ashy above, overlaid and much streaked with rusty red, which becomes bright bay on rump, tail and wings; white below with large arrow-shaped spots and streaks, numerous on breast; feet stout, with long claws; L. 7; W. 3½; T. 3. E. N. Am.; migrating early; one of the handsomest streaked sparrows.

19. ***EUSPIZA,*** Bonaparte. Black-Throated Buntings.

1. ***E. americana,*** (Gm.) Bon. Black-Throated Bunting. Grayish and streaked above; wing coverts chestnut; line over eye, maxillary stripe, edge of wing, breast and part of belly yellow; throat patch black; otherwise white below; ♀ with little chestnut, and the black reduced to a few streaks; L. 6¾; W. 3¼; T. 2¾. Meadows, etc., Conn. to Kansas, chiefly westward; a handsome bird with sleek plumage, and a peculiar, but scarcely musical song.

2. ***E. townsendi,*** (Aud.) Bon. Townsend's Bunting. Upper parts, head, neck, etc., slaty blue; no chestnut, and little yellow or black. Smaller, a doubtful species. Only one specimen known from E. Penn.

20. ***GONIAPHEA,*** Bowdich. Black-Headed Grosbeaks. = *Hedymeles*, Cabanis.

1. ***G. ludoviciana,*** (L.) Bowdich. Rose-Breasted Grosbeak. ♂ with head, neck and upper parts mostly black, with white on rump, wings and tail; belly white;

breast and under wing coverts of an exquisite rose-red; bill very stout, pale; ♀ olive brown, much streaked, with the under wing coverts saffron yellow; head with whitish bands; L. 8½; W. 4; T. 3¼. E. U. S., abundant; perhaps our handsomest bird, and one of our most brilliant songsters.

21. *GUIRACA,* Swainson. BLUE GROSBEAKS.

1. ***G. cœrulea,*** (L.) Sw. BLUE GROSBEAK. ♂ rich blue; feathers about bill, wings and tail, black; wing bars chestnut; ♀ yellowish brown, with whitish wing bars; L. 7; W. 3½; T. 2¾. Southern, N. to N. Y. and Wis.; a fine songster.

22. *CYANOSPIZA,* Baird. INDIGO BIRDS.

1. ***C. cyanea,*** (L.) Baird. INDIGO BIRD. ♂ Indigo blue, clear on head, greenish behind; ♀ plain warm brown, obscurely streaky, known from other small sparrows by a dusky line along the gonys; L. 5¾; W. 3; T. 2¾. E. U. S., abundant in summer; a tireless songster.

2. ***C. ciris,*** (L.) Baird. NONPAREIL. PAINTED BUNTING. ♂ head and neck blue; under parts, etc., vermillion; shoulders, etc., green; ♀ green, yellowish below; L. 5½. Southern, N. to S. Ills. (*Nelson.*)

23. *CARDINALIS,* Bonaparte. CARDINAL GROSBEAKS.

1. ***C. virginianus,*** (Brisson) Bon. CARDINAL GROSBEAK. RED BIRD. Clear red, ashy on back; chin and forehead black; crest conspicuous; ♀ ashy brown, more or less washed with red; L. 8½; W. 4; T. 4½. E. U. S., southerly, N. to Mass. and N. Wis.; abundant. A brilliant songster, much sought as a cage bird.

24. *PIPILO,* Vieillot. TOWHEE BUNTINGS.

1. ***P. erythrophthalmus,*** (L.) Vieill. CHEWINK. MARSH

Robin. Black, belly white; sides chestnut; outer tail feathers, primaries, and inner secondaries with white; ♀ clear brown instead of black; L. 8½; W. 3½; T. 4. E. U. S., abundant every where.

FAMILY XXXIV.—ICTERIDÆ.

(*The Orioles.*)

Primaries 9; bill with the commissure angulated, as in *Fringillidæ*, but usually lengthened, rarely shorter than head, straight or gently curved, without notch or rictal bristles; culmen usually extending up on the forehead, dividing the frontal feathers. Legs stout, tarsus strictly oscine. Plumage usually brilliant or lustrous, predominant color generally black, often with red or yellow; females usually different, smaller in size, brown or streaky in the lustrous species, and yellowish or dusky in the brightly colored ones. Notes usually sharp, often richly melodious, in other cases harsh.

Genera about twenty, species one hundred, all American, some of the short-billed forms scarcely distinct from *Fringillidæ;* others are as closely related to *Sturnidæ* (Old World Starlings) and *Corvidæ*. There are three sub-families, of which *Agelæinæ* includes most of our species. *Icterinæ* includes *Icterus*, while *Scolecophagus* and *Quiscalus* belong to *Quiscalinæ*.

I. Tail feathers rigid, acute; middle toe and claw longer than tarsus; black and whitish (♂) or brownish, streaked (♀); bill short, finch-like. . . . . . Dolichonyx, 1.

II. Feathers of crown bristle-tipped; tail short, its feathers acute; yellow below, a black breast patch; bill long. Sturnella, 5.

III. Lateral claws elongated; black or brown, yellow on head and neck; length more than 8. . . Xanthocephalus, 4.

IV. With none of the above combinations of characters.

* Length at least more than 7.

† Bill horn-blue, very acute; black or olivaceous, with orange or yellow. . . . . ICTERUS, 6.

†† Bill blackish; plumage every where streaked; usually a rusty tinge on throat and bend of wing.
♀ of AGELÆUS, 3.

††† Bill jet black; plumage in ♂ black, in ♀ duller, streaky, or plain brown.

‡ Glossy black; bend of wing red, bordered by buffy and whitish. . . . . . ♂ of AGELÆUS, 3.

‡‡ Black; head and neck rich lustrous brown.
♂ of MOLOTHRUS, 2.

‡‡‡ Iridescent black throughout; wings scarcely longer than tail; length more than 10. . QUISCALUS, 8.

‡‡‡‡ Black, often obscured by brownish or rusty; no red or yellow; wings longer than tail; length 9 to 10.
SCOLECOPHAGUS, 7.

** Length less than 7.

a. Dusky gray brown; bill blackish, shortened, finch-like.
♀ of MOLOTHRUS, 2.

aa. Black with chestnut or orange (♂), or else olive and yellowish (♀); bill acute, bluish or brown. . . ICTERUS, 6.

1. DOLICHONYX, Swainson. BOBOLINKS.

1. ***D. oryzivorus,*** (L.) Sw. BOBOLINK. REED BIRD. RICE BIRD. ♂ in Spring black, neck buffy, shoulders and rump ashy white, back streaky; ♀ and fall ♂ yellowish brown, streaked above,—dull yellow birds, resembling sparrows but known by the acute tail feathers; L. 7½; W. 4; T. 3. E. U. S., abundant in meadows northward, where, in the breeding season, it is our merriest and most delightful songster. Retiring southward in the fall, it fattens in the rice swamps and becomes a "game bird."

2. MOLOTHRUS, Swainson. COW BIRDS.

1. ***M. ater,*** (Bodd.) Gray. COW BIRD. ♂ iridescent black, head and neck glossy brown; ♀ much smaller,

dusky brown; L. (♂) 8; W. 4; T. 3. U. S., abundant; noted for its parasitic habits. [*M. pecoris* (Gmel.), Sw.]

3. AGELÆUS, Vieillot. Red-Wing Black Birds.

1. ***A. phœniceus,*** (L.) V. Red-Winged Starling. Swamp Black Bird. ♂ glossy (not iridescent) black, lesser wing covers scarlet, with buffy and paler edgings; ♀ dusky, streaked; L. 9; W. 5; T. 4. U. S., every where abundant.

4. XANTHOCEPHALUS, Bonaparte. Yellow-Headed Black Birds.

1. ***X. icterocephalus,*** (Bon.) Baird. Yellow-Headed Black Bird. ♂ black with white wing patch; head and neck rich yellow; ♀ smaller, browner, with less yellow; L. 10; W. 5½; T. 4½. Southwestern, E. to L. Michigan.

5. STURNELLA, Vieillot. Meadow Larks.

1. ***S. magna,*** (L.) Sw. Meadow Lark. Brownish and much streaked above; chiefly yellow below, a black crescent on breast. L. 10; W. 5; T. 3½. U. S.; very abundant. (*S. neglecta*, Aud., is the Western variety, Illinois S. and W., with "a much sweeter song," and some slight differences of plumage.)

6. ICTERUS, Brisson. American Orioles.

1. ***I. baltimore,*** (L.) Daudin. Baltimore Oriole. Golden Robin. Fire Bird. Black; bend of wing, rump, most tail feathers, and under parts from the breast orange of varying intensity; ♀ duller, olivaceous and yellow; L. 7¾; W. 3⅝; T. 3. E. U. S., abundant; noted for its elaborate hanging nest.

2. ***I. spurius,*** (L.) Bon. Orchard Oriole. ♂ black; rump, bend of wing and lower parts deep chestnut; ♀

yellowish olive, quite small; young yellow, with various black or chestnut traces; L. 7; W. 3½; T. 3. E. U. S., rather southerly.

7. SCOLECOPHAGUS, Swainson. RUSTY BLACK BIRDS.

1. ***S. ferrugineus,*** (Gm.) Sw. RUSTY GRACKLE. RUSTY BLACK BIRD. ♂ glossy black and rusty in autumn; ♀ dusky, lustreless; bill slender; L. 9½; W. 4¾; T. 4. E. U. S.

2. ***S. cyanocephalus,*** (Wagl.) Cab. BREWER'S BLACK BIRD. ♂ black with green lustre, head glossed with purple; ♀ dusky; L. 10; W. 5⅛; T. 4½. W., E. to Ills. and Wis.

8. QUISCALUS, Vieillot. CROW BLACK BIRDS.

1. ***Q. purpureus,*** (Bartr.) Licht. CROW BLACK BIRD. PURPLE GRACKLE. Iridescent black, lustre on head purplish, on body bronzy; L. 13; W. 5½; T. 5⅛. E. U. S., abundant.

FAMILY XXXV.—CORVIDÆ.

(The Crows and Jays.)

Primaries 10; first about half length of second; nostrils usually concealed by tufts of bristly feathers, which are branched to their tips. Bill long and strong, usually notched, commissure not angulated. Tarsus oscine, its sides undivided and separated from the scutella in front by a groove which is either naked or filled in with small scales. Voice usually harsh and unmusical.

Birds of large size, the largest of the *Oscines*, found almost every where. Genera about forty; species one hundred and seventy-five. Our two sub-families, *Corvinæ*, the Crows, and *Garrulinæ* the Jays, are usually readily distinguishable.

* Tail much shorter than the long, pointed wings. (*Corvinæ.*)

† Plumage glossy black. . . . . . . CORVUS, 1.

** Tail longer than the short, rounded wings. (*Garrulinæ.*)

‡ Conspicuously crested; chiefly blue; quills black barred. CYANURUS, 3.

‡‡ Iridescent black and white; tail much longer than wings. PICA, 2.

‡‡‡ Chiefly gray, no blue; tail scarcely longer than wings. PERISOREUS, 4.

1. CORVUS, Linnæus. RAVENS.

1. ***C. corax,*** L. RAVEN. Feathers of throat stiffened, elongated, narrow and lanceolate, their outlines very distinct; L. 25; W. 17; T. 10. N. Am., chiefly north and westward; rare E. of the Mississippi. Also European. (*C. carnivorus*, Bartr.)

2. ***C. americanus,*** Aud. CROW. Feathers of throat short, broad, obtuse, with their webs blended; gloss of plumage purplish violet; head and neck scarcely lustrous; L. 20; W. 13; T. 8. E. N. Am., chiefly eastward; abundant. (*C. frugivorus*, Bartr.)

3. ***C. ossifragus,*** Wilson. FISH CROW. Gloss of plumage green and violet, evident on head and neck; L. 16; W. 11; T. 7. New England to Florida, chiefly southern, and found only along the coast. (*C. maritimus*, Bartr.)

2. PICA, Cuvier. MAGPIES.

1. ***P. melanoleuca*** (Vieill.) var. ***hudsonica,*** (Sab.) Coues. MAGPIE. Lustrous black; belly, shoulders, and wing-edgings white; L. 19; W. 8½; T. 13, much graduated. Western, E. to L. Michigan.

3. CYANURUS, Swainson. BLUE JAYS.

1. ***C. cristatus,*** (L.) Sw. BLUE JAY. Blue; collar and frontlet black; grayish below; wings and tail clear

blue-barred; outer tail feathers and secondaries tipped with white; L. 12; W. 5½; T. 5¾. N. E. Am., abundant.

4. *PERISOREUS*, Bonaparte. GRAY JAYS.

1. ***P. canadensis***, (L.) Bon. CANADA JAY. WHISKEY JACK. Ashy gray with blackish and whitish markings; L. 10¾; W. 5¾; T. 6. Northern, S. to New England in Winter.

FAMILY XXXVI.—TYRANNIDÆ.

(*The Flycatchers.*)

Primaries 10; first more than ¾ length of second, and one or more of them often attenuate; bill broad, triangular, depressed, abruptly hooked and notched at tip, with long rictal bristles; commissure nearly straight; nostrils small, usually partly concealed. Tarsus "clamatorial," the scutella extending around its back. Feet small, for perching. Mouth capacious; notes simple, often pleasant; changes of plumage slight; ours mostly olivaceous.

A large family of eighty genera, and more than three hundred species; all American and mostly tropical. All are insectivorous, most of them pre-eminently so; they are, therefore, in our latitude, migratory.

* First primaries evidently attenuate; crown with concealed bright red or yellow crest (in adult).

† Tail widely forked, about twice as long as wings. MILVULUS, 1.

†† Tail nearly even, shorter than wings. . . TYRANNUS, 2.

** First primaries not obviously attenuate; crown plain, sometimes crested.

‡ Wings edged with chestnut, not much longer than tail; length 8 or more. . . . . MYIARCHUS, 3.

‡‡ Wings not chestnut-edged, not much longer than tail; tarsus longer than middle toe and claw; bill black; length 6½ to 7½. . . . . SAYORNIS, 4.

‡‡‡ Wings longer than tail; tarsus shorter than middle toe and claw; bill not all black; length 6 or more. CONTOPUS, 5.

‡‡‡‡ Wings not much longer than tail; middle toe and claw not longer than tarsus; bill mostly pale below; length $3\frac{1}{4}$ or less. . . . . EMPIDONAX, 6.

1. MILVULUS, Swainson. FORK-TAILED FLYCATCHERS.

1. ***M. forficatus,*** (Gm.) Sw. SCISSOR-TAIL. Ashy, tail, shoulders, sides, etc., with much red; L. 13; W. 5; T. 8. S. W., N. to Kansas, straying to New Jersey.

2. ***M. tyrannus,*** (L.) Bon. FORK-TAILED FLYCATCHER. Larger, no red, tail still more elongate. Tropical, straying to N. J. and La.

2. TYRANNUS, Cuvier. KING BIRDS.

1. ***T. carolinensis,*** (L.) Baird. KING BIRD. BEE MARTIN. Blackish ash, white below; tail black, white-tipped; L. $8\frac{1}{2}$; W. $4\frac{5}{8}$; T. $3\frac{1}{2}$. U. S., chiefly eastward; abundant. "Destroys a thousand noxious insects for every bee it eats!" (*Coues.*)

2. ***T. verticalis,*** Say. ARKANSAS FLYCATCHER. Belly yellow; tail white-edged. Western, straying to N. J.

3. MYIARCHUS, Cabanis. CRESTED FLYCATCHERS.

1. ***M. crinitus,*** (L.) Cab. GREAT CRESTED FLYCATCHER. Scarcely crested; olivaceous, yellow below, with bright chestnut on wings and tail; L. $8\frac{3}{4}$; W. 4; T. 4. E. U. S., chiefly southerly, N. to N. Wis. A handsome bird, "noted for the habitual use of cast-off snake skins in the structure of its nest."

4. SAYORNIS, Bonaparte. PEWEES.

1. ***S. fuscus,*** (Gm.) Baird. PEWEE. PHŒBE. PEWIT. Olive brown, head and tail darker; yellow below, more

or less; L. 7; W. $3\frac{1}{3}$; T. $3\frac{1}{4}$. E. U. S., abundant; known by its black bill.

5. *CONTOPUS,* Cabanis. Wood Pewees.

1. ***C. borealis,*** (Sw.) Baird. Olive-Sided Flycatcher. Rictal bristles short, one-fourth length of bill; tuft of white cottony feathers on sides very conspicuous; middle line of belly distinctly and abruptly white; otherwise olive brown, paler or yellowish below; L. $7\frac{1}{2}$; W. $4\frac{1}{3}$; T. 3. Northern, S. to N. Y.

2. ***C. virens,*** (L.) Cab. Wood Pewee. Rictal bristles half length of bill; cottony tuft inconspicuous; wing bands whitish or rusty; olive brown above; pale or yellowish below; lower mandible usually pale; L. $6\frac{1}{4}$; W. $3\frac{1}{2}$; T. 3. U. S., very abundant.

3. ***C. richardsoni,*** (Sw.) Bd. Western Wood Pewee. Darker; bill dusky below. N. W., E. to Wis.; nearly like the preceding, but the notes and nesting different.

6. *EMPIDONAX,* Cabanis. Least Flycatchers.

1. ***E. acadicus,*** (Gm.) Baird. Small Green-Crested Flycatcher. Clear olive green, wing bands buffy; whitish becoming yellowish below; yellowish ring about eyes; bill pale below; primaries nearly an inch longer than secondaries; 2d, 3d and 4th primaries nearly equal, and much longer than 1st and 5th; 1st much longer than 6th; L. 6; W. 3; T. $2\frac{3}{4}$; Ts. $\frac{2}{3}$; Tcl. $\frac{1}{2}$. E. U. S., frequent.

2. ***E. traillii,*** (Aud.) Baird. Traill's Flycatcher. Olive brown, duller than preceding; bill pale below; 5th primary about as long as 4th, 1st not much longer than 6th; middle toe $\frac{2}{3}$ length of tarsus; longest primary $\frac{2}{3}$ inch longer than secondaries; L. $5\frac{3}{4}$; W. $2\frac{3}{4}$; T. $2\frac{1}{2}$; Ts. $\frac{2}{3}$; Tcl. $\frac{3}{6}$. U. S.

3. ***E. minimus,*** Baird. Least Flycatcher. Olive gray; bill blackish below; wings like preceding, but longest primary but $\frac{1}{2}$ inch longer than secondaries; middle toe half as long as tarsus; bill less than $\frac{1}{2}$ inch; L. 5; W. $2\frac{1}{2}$; T. $2\frac{1}{4}$. E. N. Am., abundant.

4. ***E. flaviventris,*** Baird. Yellow-Bellied Flycatcher. Clear olive green; yellow below, becoming bright yellow (not merely yellowish as in the others) on the belly; first primary about equal to sixth; feet as in *acadicus;* bill yellow below; L. $5\frac{1}{4}$; W. $2\frac{3}{4}$; T. $2\frac{1}{2}$. E. U. S.

ORDER H.—PICARIÆ.

(*Picarian Birds.*)

Hind toe small, sometimes wanting, occasionally elevated; its claw shorter than that of middle toe (with rare exceptions); 3d and 4th toes often with less than the normal number of joints; 2d and 4th toes sometimes versatile. Wing coverts larger and in more numerous series than in the *Passeres.* Primaries 10, first rarely short; tail feathers 10 (8 to 12). Musical apparatus imperfect. Sternum non-passerine. Tarsus never oscine. Nature altricial. A highly diversified group, the members of which have little in common except their want of resemblance to other birds.

FAMILY XXXVII.—CAPRIMULGIDÆ.

(*The Goatsuckers.*)

Bill very short, "fissirostral," the gape exceedingly deep and wide, reaching to below the eyes, and usually with prominent rictal bristles. Wings long and pointed; secondaries lengthened. Plumage long and loose, owl-like. Tail feathers 10. Feet very small; tarsus short, partly feathered; toes slightly webbed at base, the hind toe somewhat elevated. Genera fourteen; species one hundred or more, widely diffused; chiefly insectivorous.

* Tail rounded; rictal bristles very long. . ANTROSTOMUS, 1.

** Tail forked; rictal bristles inconspicuous. . CHORDEILES, 2.

1. *ANTROSTOMUS,* Gould. WHIPPOORWILLS.

1. ***A. vociferus,*** (Wils.) Bon. WHIPPOORWILL. NIGHT JAR. Grayish, much variegated; pectoral bar and ends of outer tail feathers white (♂) or tawny (♀); rictal bristles unbranched; L. 10; W. 6; T. 5. E. U. S., abundant, nocturnal; noted for its "solemn and prophetic" cry.

2. ***A. carolinensis,*** (Gm.) Gould. CHUCKWILL'S WIDOW. More reddish; rictal bristles with lateral filaments; L. 12; W. 9; T. 6½. Southern, N. to Ills. (*Nelson.*)

2. *CHORDEILES,* Swainson. NIGHT HAWKS.

1. ***C. virginianus,*** (Gm.) Bon. NIGHT HAWK. BULL BAT. Blackish, variegated; a large wing spot, bar across tail, and V-shaped blotch on throat—white in ♂, tawny or obscure in ♀; L. 9½; W. 8; T. 5. U. S.; abundant. [*C. popetue*, (Vieill.) Bd.]

FAMILY XXXVIII.—CYPSELIDÆ.

(*The Swifts.*)

Bill fissirostral, as in *Caprimulgidæ* and *Hirundinidæ.* Wings very long, thin and pointed; secondaries very short. Feet small, weak; hind toe often elevated or otherwise turned; toes completely cleft. No rictal bristles. Tail feathers 10; plumage compact. In most species the salivary glands are highly developed, and their secretion is used as a glue in the construction of the nest; species of *Collocalia* thus form the edible bird's nest. Small birds of the warmer parts of the world, bearing a superficial resemblance to Swallows, but structurally very different, being closely related to the Humming Birds. Genera six or eight; species about fifty.

* Tarsus bare, longer than middle toe; tail feathers with the shafts spinous, projecting beyond the plumage. CHÆTURA, 1.

1. CHÆTURA, Stephens. CHIMNEY SWALLOWS.

1. ***C. pelagica,*** (L.) Baird. CHIMNEY SWIFT. Sooty brown; throat paler; L. 5¼; W. 5; T. 2. E. U. S., abundant.

FAMILY XXXIX.—TROCHILIDÆ.

(*The Humming Birds.*)

Bill subulate, usually longer than the head, straight or curved; tongue capable of great protrusion. Wings long and pointed, the secondaries short, only six in number; tail of ten feathers. Feet very small, with sharp claws. Smallest of all birds and among the most brilliantly colored. Genera seventy-five; species three hundred or more, thus forming one of the largest families in Ornithology. All are American, and most of them tropical, but our common species ranges far into British America.

* First primary not attenuate, bowed or curved inwards. TROCHILUS, 1.

1. TROCHILUS, Linnæus. RUBY-THROATED HUMMING BIRDS.

1. ***T. colubris,*** L. RUBY-THROATED HUMMING BIRD. ♂ metallic green above; a ruby-red gorget; tail deeply forked, uniform purplish; ♀ without red, the tail vari-

egated; L. 3¼; W. 1⅔; T. 1¼; B. ⅔. E. N. Am.; abundant in summer.

FAMILY XL.—ALCEDINIDÆ.

(*The Kingfishers.*)

Head large; bill long, straight and strong, usually longer than head; gape deep, tomia not serrate. Wings long; tail short. Legs quite small; feet syndactyle—the outer and middle toes united to their middle, a continuous sole beneath; tibia naked below. Tail feathers twelve. Species about one hundred, chiefly of the tropical parts of the Old World and Australia. Many of them feed upon fishes, and nearly all are remarkable for their brilliant coloration.

* Head crested. . . . . CERYLE, 1.

1. CERYLE, Boie. KINGFISHERS.

> *Ispida*, Swainson.

1. ***C. alcyon,*** (L.) Boie. BELTED KINGFISHER. Ashy blue above, a bluish band across breast; white below; ♀ with sides and band across belly chestnut; tail barred with white; L. 13; W. 6; T. 3½; B. 2, or more. N. Am.; every where.

FAMILY XLI.—CUCULIDÆ.

(*The Cuckoos.*)

Bill compressed, lengthened, decurved; usually without rictal bristles or nasal tufts. Tail long and soft, of eight to twelve feathers. Tongue not extensible. Feet zygodactyle, by reversion of fourth toe. Species about two hundred, in various parts of the world.

* Plumage lustrous olive gray or drab; arboreal. COCCYGUS, 1.

1. COCCYGUS, Vieillot. AMERICAN CUCKOOS.

1. ***C. americanus,*** (L.) Bon. YELLOW-BILLED CUCKOO. Bill yellow below; wings with much cinnamon red; middle tail feathers like the back; outer ones black with broad white tips; L. 12; W. 5½; T. 6. U. S.

2. ***C. erythrophthalmus,*** (Wils.) Baird. BLACK-BILLED CUCKOO. Bill chiefly black; wings with little or no reddish; tail feathers all brownish, obscurely whitish at tips; L. 11½; W. 5; T. 6¼. E. U. S.

FAMILY XLII.—PICIDÆ.

(*The Woodpeckers.*)

Bill stout, usually straight, with the tip truncate or acute, fitted for hammering or boring into wood. Tongue long, flattish, barbed, capable of great protrusion, adapted for securing insects (except in *Sphyrapicus*); hyoid apparatus peculiar, its horns generally quite long, curving around the skull behind. Feet zygodactyle, outer toe permanently reversed; hind toe present (except in *Picoides*); claws compressed, sharp and strong. Tail feathers 12, rigid and acuminate, outer pair short, concealed; tail never forked; nasal tufts usually present.

Chiefly arboreal; all (except *Sphyrapicus*, which is truly a "Sap-Sucker,") are pre-eminently insectivorous and hence they are of the greatest service to the farmer. Voice loud and often harsh. Colors generally bright, the male at least having almost always red on the head; sexes usually slightly different. Species two hundred and fifty, abundant almost every where.

* Conspicuously crested; length 18 or more.
— Bill and nasal feathers dark. . HYLOTOMUS, 1.
— Bill and nasal feathers pale. . CAMPEPHILUS, 2.
** Not crested; toes 3 only, hallux wanting. . PICOIDES, 4.
*** Not crested; toes 4, length less than 14.

† Tongue obtuse, brushy; ridges on upper mandible running into the tomia; belly with some yellow. SPHYRAPICUS, 5.

†† Tongue acute, barbed; ridges on sides of upper mandible reaching the tip; no yellow; quills (in ours) with round white spots. . . . . PICUS, 3.

††† Tongue acute, barbed; ridges on sides of upper mandible wanting or indistinct.

‡ Back barred.

a. Belly with round black spots; feathers of wings and tail yellow or orange beneath. . . COLAPTES, 8.

aa. Belly unspotted, tinged with red or yellow; no yellow on quills. . . . . CENTURUS, 6.

‡‡ Back not barred; body lustrous blue-black; rump, secondaries, and under parts white; head and neck red in adults, grayish in young. . . . MELANERPES, 7.

1. HYLOTOMUS, Baird. BLACK WOODCOCKS.

1. ***H. pileatus,*** (L.) Baird. PILEATED WOODPECKER. LOGCOCK. Black; white streak down neck; crest and cheek patch scarlet in ♂; cheeks and front of crest black in ♀, L. 18; W. 9½; T. 7. N. Am.; in heavy timber.

2. CAMPEPHILUS, Gray. IVORY-BILLED WOODPECKERS.

1. ***C. principalis,*** (L.) Gray. GREAT IVORY-BILLED WOODPECKER. Black with white markings; crest scarlet in ♂, black in ♀; L. 21; W. 11; T. 8. Southern, N. to S. Ills.

3. PICUS, Linnæus. SPOTTED WOODPECKERS.

1. ***P. borealis,*** Vieill. RED-COCKADED WOODPECKER. Black and white, spotted and crosswise banded, but not streaked; a red line on each side of head in ♂; L. 8½; W. 4½; T. 3½. Southern States in swamps, N. to Penn.

2. ***P. villosus,*** L. HAIRY WOODPECKER. BIG SAP-SUCKER. Spotted and lengthwise streaked, but not

banded; back black with a long white stripe; outer tail feathers wholly white; L. 9; W. 5; T. 3½; a scarlet nuchal band in ♂ only. U. S.; every where.

3. ***P. pubescens,*** L. Downy Woodpecker. Little Sap-Sucker. Much smaller; outer tail feathers black and white, barred, otherwise precisely like the other; L. 6½; W. 3¾; T. 2¾. U. S.; every where.

4. PICOIDES, Lacepede. Three-Toed Woodpeckers.

1. ***P. arcticus,*** (Sw.) Gray. Black-Backed Woodpecker. Black and white; crown yellow in ♂, plain in ♀; back uniform black; L. 9; W. 5; T. 3⅝. Northern, S. to U. S. in winter.

2. ***P. americanus,*** Brehm. Banded Three-Toed Woodpecker. Back with a white lengthwise stripe; otherwise as above; L. 8; W. 4½; T. 3½. Arctic, S. in winter to New England.

5. SPHYRAPICUS, Baird. Sap-Sucking Woodpeckers.

1. ***S. varius,*** (L.) Baird. Yellow-Bellied Woodpecker. Black and white above; black on breast; chiefly yellowish below; white wing patch; crown red in adult, chin scarlet in ♂; L. 8¼; W. 4¾; T. 3⅓. U. S., abundant.

6. CENTURUS, Swainson. Red-Bellied Woodpeckers.

1. ***C. carolinus,*** (L.) Bon. Red-Bellied Woodpecker. Grayish, much barred above with black and white; crown and nape crimson in ♂, crown ashy in ♀, belly reddish-tinged; L. 9¾; W. 5; T. 3½. E. U. S., rather southerly; N. to N. Wis.

7. MELANERPES, Swainson. Red-Headed Woodpeckers.

1. ***M. erythrocephalus,*** (L.) Sw. Red-Headed Wood-

PECKER. L. 9; W. 5½; T. 3½. U. S., E. of the Rocky Mts.; abundant.

8. *COLAPTES*, Swainson. FLICKERS.

1. ***C. auratus*,** (L.) Sw. GOLDEN-WINGED WOODPECKER. HIGH-HOLER. YARUP. Head ashy, with red nuchal crescent; back olivaceous, barred with black; rump white; below pinkish brown shading into yellowish, a black crescent on breast and numerous round black spots; shafts and under surfaces of quills golden yellow; ♂ with a black maxillary patch; L. 12½; W. 6; T. 4½. E. U. S., abundant.

Var. ***mexicanus*,** (Sw.) Snow. RED-SHAFTED FLICKER. Quills with orange red instead of golden; maxillary patches in ♂ red instead of black; no nuchal crescent; no yellowish on belly. Western, E. to Kas., etc. Runs into the preceding, through *C. hybridus*. Baird.

ORDER I.—PSITTACI.

(The Parrots.)

Bill enormously thick, cered at base and strongly hooked. Feet zygodactyle by reversion of outer toe, tarsus reticulate. Tongue short, fleshy; upper jaw unusually movable. Altricial. Plumage often brilliant. In all warm regions; species three hundred and fifty-four, nearly half of which are American.

FAMILY XLIII.—ARIDÆ.

(The Macaws.)

Parrots with the head not crested, and the tail long, wedge-shaped or graduated. (*Baird.*)

* Culmen rounded; face entirely feathered except a curve about the eye; tail shorter than wings. . . CONURUS, 1.

1. *CONURUS*, Kuhl. Parroquets.

1. ***C. carolinensis***, (L.) Kuhl. Carolina Parroquet. Green; head and neck yellow; face red; wings with blue and yellow; bill white; L. 13; W. 7½; T. 6. Southwestern, formerly N. to the Great Lakes; now nearly exterminated.

ORDER J.—RAPTORES.

(*The Birds of Prey.*)

Bill powerful, cered at base, strongly hooked at the end. Feet never zygodactyle; fourth toe sometimes versatile; claws long and sharp; hind toe well developed, rarely elevated; tibia, and often tarsus, feathered. Primaries 10; tail feathers 12 (with rare exceptions). Altricial, but young downy at birth. Carnivorous birds, generally of large size and great strength, found in every part of the world.

FAMILY XLIV.—STRIGIDÆ.

(*The Owls.*)

Head very large, shortened lengthwise and greatly expanded laterally; the eyes directed forwards and partly surrounded by a disk of radiating feathers of peculiar texture; loral feathers antrorse, long and dense; feathers on the sides of forehead often elongated into ear-like tufts. Plumage very soft and lax, rendering the flight almost noiseless; its colors blended and mottled so as to render minute description difficult. External ear very large, often provided with a movable flap. Outer toe versatile; claws very sharp, long and strong. Eggs nearly spherical, pure white. Chiefly nocturnal. Sexes colored alike, ♀ usually the larger. Owls are found in every part of the globe, and most of the species have a wide range. Their habits are so well known that I need not dwell upon them here. Genera about forty; species one hundred and fifty.

* Tarsus naked or scant-feathered, facial disk perfect; no ear-tufts; middle claw pectinate; iris black. . STRIX, 1.

** Tarsus fully feathered.

† Head with evident "ear-tufts;" iris yellow.

‡ Tail about $\frac{2}{3}$ of wing; bill blackish; length more than 18. BUBO, 6.

‡‡ Tail about half length of wing; length less than 18.

a. Bill pale; length less than 12. . . . SCOPS, 5.

aa. Bill dark; length more than 12. . . . OTUS, 2.

†† Head without evident "ear-tufts."

b. Tail about $\frac{1}{2}$ length of wing; iris yellow; length less than 12. . . . . NYCTALE, 4.

bb. Tail about $\frac{2}{3}$ of wing; length 18 or more.

c. Pure white, with dark markings; toes concealed by long feathers; facial disk incomplete; bill black; iris yellow. NYCTEA, 7.

cc. Grayish, much barred; facial disk complete; bill yellow; iris black or yellow. . . . . . SYRNIUM, 3.

bbb. Tail about $\frac{3}{4}$ of wing; bill yellow; iris yellow; length about 16. . . . . SURNIA, 8.

*** Tarsus long, sparsely bristly; facial disk imperfect; middle claw simple. . . . . SPEOTYTO, 9.

1. ***STRIX,*** Linnæus. BARN OWLS.

1. ***S. flammea,*** (L.) var. ***pratincola,*** (Bon.) Ridg. BARN OWL. Face elongated; reddish or tawny, much variegated; L. 17; W. 13; T. $5\frac{1}{2}$. U. S., rather southerly.

2. ***OTUS,*** Cuvier. EARED OWLS.

1. ***O. vulgaris,*** (L.) var. ***wilsonianus,*** (Less.) Allen. LONG-EARED OWL. Ear tufts well developed, of 8 to

12 feathers; outer primary only emarginate; much variegated; L. 15; W. 12; T. 6. U. S.

2. ***O. brachyotus,*** (Gm.) Steph. SHORT-EARED OWL. Ear tufts small and inconspicuous; two outer primaries emarginate; L. 15; W. 12; T. 6. U. S. and Europe. (*Brachyotus palustris*, Auct.)

3. *SYRNIUM,* Savigny. BARRED OWLS.

* Iris black; 5 outer primaries emarginate. (*Syrnium.*)

1. ***S. nebulosum,*** (Forst.) Boie. BARRED OWL. Toes not concealed; olive brown, barred with white above; breast barred; belly streaked; L. 18; W. 14; T. 9. E. N. Am., common.

** Iris yellow; 6 outer primaries emarginate. (*Scotiaptex.*)

2. ***S. cinereum,*** (Gmel.) Aud. GREAT GRAY OWL. Toes concealed by long feathers; cinereous brown above, waved with white; breast streaked, belly barred; largest of all our owls; L. 30; W. 18; T. 12. Northern, S. in winter to N. States.

4. *NYCTALE,* Brehm. SPARROW OWLS.

1. ***N. tengmalmii,*** Gm. var. ***richardsonii,*** (Bon.) Ridg. TENGMALM'S OWL. Nostrils sunken, elongated, opening laterally; tail more than half wing; general color chocolate brown, variegated; L. 10; W. 7¼; T. 4½. Northern, S. to N. U. S.

2. ***N. acadica,*** (Gm.) Bon. SAW-WHET OWL. Nostrils prominent, nearly circular, opening anteriorly; L. 8; W. 5¾; T. 2¾. U. S., rather northerly.

5. *SCOPS,* Savigny. SCREECH OWLS.

1. ***S. asio,*** (L.) Bon. SCREECH OWL. RED OWL. Grayish, speckled and barred, or else with the grayish replaced by bright reddish; these two different styles of

plumage bearing no relation to age, sex or season; L. 10; W. 7; T. 3½. U. S., abundant.

6. *BUBO,* Dumeril. GREAT HORNED OWLS.

1. ***B. virginianus,*** (Gm.) Bon. GREAT HORNED OWL. Black, gray and buffy, variously mottled and barred; usually a whitish half-collar; ear tufts large, their feathers mostly black; L. 22; W. 16; T. 10. U. S., abundant; one of the strongest and most untamable of the Owls.

7. *NYCTEA,* Stephens. GREAT SNOW OWLS.

1. ***N. scandiaca,*** (L.) Newt. SNOWY OWL. Pure white, more or less barred with blackish; L. 23; W. 17; T. 10. Northern, S. in Winter; one of the handsomest of Owls. (*N. nivea*, Auct.)

8. *SURNIA,* Dumeril. HAWK OWLS.

1. ***S. ulula,*** (L.) Bon., var. ***hudsonica,*** (Gm.) Ridg. HAWK OWL. DAY OWL. Brown, much speckled and barred; L. 16; W. 9; T. 7. Northern, S. to Wis. and Mass.

9. *SPEOTYTO,* Gloger. BURROWING OWLS.

1. ***S. cunicularia,*** (Mol.) var. ***hypogæa,*** (Bon.) Coues. BURROWING OWL. Brownish, much spotted and variegated. L. 10; W. 7½; T. 4. Fla. and Western Plains, living in the holes of prairie dogs.

FAMILY XLV.—FALCONIDÆ.

(*The Falcons.*)

Eyes lateral, eyelids provided with lashes, usually a projecting bony eyebrow; no complete facial disk. Toes always naked, and usually tarsus also; hind toe not elevated. Head fully feathered (except in the Old World

Vulturinæ), no ear tufts. Base of stout, strongly hooked bill, not hidden by feathers. Claws very strong and sharp. Plumage usually of blended colors, barred or streaked; changes considerable; ♀ usually the larger. Genera fifty, species three hundred, abounding every where. Their habits are too well known to require description here.

I. Tarsus feathered to the toes.

a. Tarsus entirely feathered; tail 12 or more. . AQUILA, 12.

aa. Tarsus with a narrow unfeathered strip behind; tail less than 12. . . . . ARCHIBUTEO, 11.

II. Tarsus reticulate all around.

b. Upper mandible toothed; under notched; nostrils circular. FALCO, 1.

bb. Tail widely forked; outer feather twice as long as middle ones; colors black and white. . . NAUCLERUS, 3.

bbb. Claws all of same length, rounded beneath; tibial feathers close; plumage compact, without after shafts. PANDION, 2.

bbbb. Tail emarginate, and outer feather not longer than middle; head and tail white in adult. . . . ELANUS, 5.

III. Tarsus scutellate in front only (occasionally "booted.")

c. Toes not webbed at all; neck feathers lanceolate, white in adult. . . . . HALIAETUS, 13.

cc. Toes somewhat webbed at base.

d. Nostrils circular; tail less than ⅔ length of wing. ICTINIA, 4.

dd. Nostrils oval; tail more than ⅔ length of wing.

e. Tarsus feathered about half way down in front, the feathers scarcely separated behind. . ASTUR, 7.

ee. Tarsus feathered less than one-third down in front, the feathers widely separated behind. . ACCIPITER, 8.

IV. Tarsus scutellate in front and behind.

f. Face with a slight ruff; tarsus twice length of middle toe; upper tail coverts white. . . . . CIRCUS, 6.

ff. No ruff; 3 or 4 outer primaries emarginate; rump not white. BUTEO, 10.

fff. No ruff; 4 primaries emarginate; tail coverts white; tail black. . . . . ASTURINA, 9.

1. *FALCO*, Linnæus. FALCONS.

* First primary only emarginate on inner web; 2d longest, 1st shorter than 4th; tarsal plates small; sexes colored alike.

† Tarsus not longer than middle toe, scarcely feathered below joint. (*Falco.*)

1. ***F. communis,*** Gm. PEREGRINE FALCON. DUCK HAWK. Blackish ash with paler waves; below whitish, barred; black cheek patches; L. 16; W. 13; T. 7. U. S., not common.

†† Tarsus longer than middle toe and claw, feathered for some distance. (*Hierofalco*, Cuv.)

2. ***F. sacer,*** Forst. GYRFALCON. Tarsus feathered half way down, with only a bare strip behind; white or ashy with dark markings; L. 24; W. 16; T. 10. Northern regions of both continents; var. *islandicus*, S. to U. S. in winter. (*F. gyrfalco*, L.)

3. ***F. mexicanus,*** Licht. LANIER FALCON. Tarsus feathered ⅓ way down; general color brown; L. 18; W. 14; T. 8. S. W., E. to Ills.

** Two primaries emarginate; tarsal plates enlarged in front, appearing like scutella.

‡ Tarsus about equal to middle toe; basal joints of toes without transverse scutella. (*Æsalon*, Kaup.)

4. ***F. columbarius,*** L. PIGEON HAWK. AMERICAN MERLIN. Ashy blue or blackish above, variegated below; L. 13; W. 8; T. 5. U. S.

‡‡ Tarsus longer than middle toe; basal joints of toes with transverse scutella. (*Tinnunculus*, Vieill.)

5. ***F. sparverius,*** L. SPARROW HAWK. RUSTY-CROWNED FALCON. Back tawny; wings bluish and black; seven black blotches about head; tail chestnut,

with a broad black band in ♂, and a narrow terminal one of white; below white or tawny; L. 11; W. 7; T. 5. U. S., abundant.

2. *PANDION,* Savigny. Ospreys.

1. ***P. haliaetus,*** (L.) Savigny. Osprey. Fish Hawk. Dark brown; head, neck and under parts mostly white; feet very large; L. 24; W. 20; T. 10. U. S.; feeds on fishes.

3. *NAUCLERUS,* Vigors. Swallow-Tailed Kites.

1. ***N. furcatus,*** (L.) Vig. Swallow-Tailed Kite. Lustrous black; head, neck and lower parts white; W. 17; T. 14. Southern, N. to Penn. and Minn.

4. *ICTINIA,* Vieillot. Blue Kites.

1. ***I. subcœruleus,*** (Bart.) Coues. Mississippi Kite. Chiefly lead blue, wings with chestnut; L. 15; W. 12; T. 6½. Southern, N. to Penn. and Wis. (*I. mississippiensis*, Auct.)

5. *ELANUS,* Savigny. White-Tailed Kites.

1. ***E. glaucus,*** (Bartr.) Coues. Black-Shouldered Kite. L. 17. Southern, N. to S. Ills. (*E. leucurus*, Auct.)

6. *CIRCUS,* Lacepede. Marsh Harriers.

1. ***C. cyaneus*** (L.) var. ***hudsonius,*** (L.) Coues. Marsh Harrier. Pale bluish or brown; rump and under parts whitish; L. 18; W. 15; T. 9. N. Am., abundant.

7. *ASTUR,* Lacepede. Goshawks.

1. ***A. palumbarius,*** (L.) var. ***atricapillus,*** (Wils.) Coues. Goshawk. Slate blue with white superciliary stripe; tail with four dark bars; L. 24; W. 14; T. 11. Northern, S. to U. S. in winter.

8. *ACCIPITER*, Brisson. HAWKS.

= *Nisus*, Cuvier.

1. ***A. fuscus***, (Gm.) Bon. SHARP-SHINNED HAWK. "PIGEON HAWK." Bare portion of tarsus in front, longer than middle toe; tarsus "booted" in ♂; general color dark brown; L. 12; W. 7; T. 6. U. S., abundant.

2. ***A. cooperi***, Bon. CHICKEN HAWK. Bare tarsus shorter than middle toe; L. 18; W. 10; T. 8. N. Am.

9. *ASTURINA*, Vieillot. GRAY HAWKS.

1. ***A. nitida***, (Lath.) var. ***plagiata***, (Schl.) Coues. GRAY HAWK. L. 18; W. 10; T. 7½. Mexican, straying to S. Ills.

10. *BUTEO*, Cuvier. BUZZARDS.

* 4 outer primaries emarginate on inner web. (*Buteo.*)

1. ***B. borealis***, (Gm.) Vieill. HEN HAWK. RED-TAILED BUZZARD. Dark brown; much barred and streaked; tail bright chestnut red above; L. 23; W. 15½; T. 8½. U. S., common.

2. ***B. lineatus***, (Gm.) Jard. RED-SHOULDERED BUZZARD. Dark reddish brown, variegated; bend of wing orange brown; L. 22; W. 14; T. 9. Smaller than the preceding, although nearly as long. E. N. Am., abundant.

** 3 outer primaries emarginate on inner web. (*Craxirex*, Gould.)

3. ***B. swainsoni***, Bon. SWAINSON'S BUZZARD. Gray, variously streaked, usually a dark area on throat and breast; tail with six or more narrow dark bars; variable; L. 20; W. 16; T. 8½. Western; E. to Ind. and Mass.

4. ***B. pennsylvanicus***, (Wils.) Bon. BROAD-WINGED HAWK. Brown above, whitish or fulvous below, variously streaked and barred; conspicuous dark cheek patches; tail with broad dark bands alternating with narrower

pale ones, white-tipped; L. 18; W. 11; T. 7. E. U. S.; a stout, handsome, though small hawk.

11. ***ARCHIBUTEO,*** Brehm. Squirrel Hawks.

1. ***A. lagopus,*** (Brunn.), var. ***sancti-johannis,*** (Gm.) Ridg. Rough-Legged Hawk. Black Hawk. Chiefly whitish but sometimes entirely black; L. 24; W. 18; T. 10. N. Am.

12. ***AQUILA,*** Möhring. Golden Eagles.

1. ***A. chrysaetus,*** (L.) Golden Eagle. Glossy purplish brown; head and neck golden brown; quills blackish; L. 36; W. 25; T. 16. N. Am., chiefly northerly.

13. ***HALIAETUS,*** Savigny. Bald Eagles.

1. ***H. leucocephalus,*** (L.) Savigny. Bald Eagle. Dark brown; head, neck and tail white (after the third year); L. 36; W. 25; T. 14. N. Am., every where; feeds on fishes. "A piratical parasite of the Osprey, otherwise notorious as the emblem of the Republic." (*Coues.*)

FAMILY XLVI.—CATHARTIDÆ.

(*The New World Vultures.*)

Head and part of neck bare. Eyes lateral, not overhung; ears small. Bill lengthened, weak and but little hooked; nostrils perforate. Wings very long and strong, giving a strength and grace of flight scarcely excelled. Hind toe short, and elevated; front toes long, somewhat webbed, with rather weak and straightish claws. Large turkey-like raptores, without the strength and spirit of the hawks and owls; "voracious and indiscriminate gormandizers of carrion and animal refuse of all sorts, hence efficient and almost indispensable scavengers in the warm countries where they abound." (*Coues.*)

Two species, the Condor and the California Vulture, are among the largest birds of flight in the world. All are American, the Old World Vultures (*Vulturinæ*) being Vulture-like hawks. Genera five; species six or eight.

* Wings very long, primaries reaching to end of tail or farther; skin of neck not corrugated; a tuft of bristles in front of eye. . . . . RHINOGRYPHUS, 1.

** Wings short, scarcely reaching middle of tail; skin of neck corrugated; no bristles in front of eye. . CATHARISTA, 2.

1. *RHINOGRYPHUS,* Ridgway. TURKEY BUZZARDS.

< *Cathartes*, Illiger.

1. ***R. aura,*** (L.) Ridg. TURKEY BUZZARD. Black, lustrous above; skin of head and neck red; L. 30; W. 22; T. 12. N. Am., abundant, southward.

2. *CATHARISTA,* Vieillot. CARRION CROWS.

1. ***C. atrata,*** (Bartr.) Gray. CARRION CROW. Uniform dull black; L. 24; W. 17; T. 8. N. C. to Mexico; rarely straying northward.

ORDER K.—COLUMBÆ.

(*The Doves.*)

Bill straight, compressed, the horny tip separated by a constriction from the soft part. Nostrils opening beneath a soft, tumid membrane. Frontal feathers sweeping in a strongly convex outline across base of upper mandible; tomiæ meeting. Hind toe on a level with the rest (except in *Starnœnas*, etc.), the others usually not webbed. Tarsus mostly scutellate in front, elsewhere reticulate, the plates soft. Head small.

Plumage soft, compact, the feathers very loosely inserted. Altricial; monogamous.

FAMILY XLVII.—COLUMBIDÆ.

(*The Doves.*)

Wings long, pointed. Tail never forked, of 12 or 14 feathers; male with the neck iridescent. Species about three hundred, found in most regions, but most abundant in the East Indies. Besides the following, quite a number of species occur in the Southern States. The common domesticated dove (*Columba livia*) is a fair type of the family.

* Tarsus feathered at the suffrago, shorter than the lateral toes; tail very long, wedge-shaped, of 12 feathers. ECTOPISTES, 1.

** Tarsus entirely bare, longer than the lateral toes.

† Tail long, pointed, of 14 feathers; length more than 10. ZENÆDURA, 2.

†† Tail short, rounded, of 12 feathers; length less than 8. CHAMÆPELIA, 3.

1. ECTOPISTES, Swainson. PASSENGER PIGEONS.

1. ***E. migratorius,*** (L.) Sw. WILD PIGEON. Bluish with reddish and violet tinges, reddish below; L. 17; W. $7\frac{1}{2}$; T. 8. N. A., abundant; gregarious.

2. ZENÆDURA, Bonaparte. MOURNING DOVES.

1. ***Z. carolinensis,*** (L.) Bon. MOURNING DOVE. TURTLE DOVE. CAROLINA DOVE. Brownish olive, glossed with blue and wine color; plumage with metallic lustre; L. 12; W. $5\frac{3}{4}$; T. $6\frac{3}{4}$. U. S., abundant.

3. CHAMÆPELIA, Swainson. GROUND DOVES.

1. ***.C passerina,*** (L.) Sw. GROUND DOVE. Grayish olive, with bluish gloss; L. $6\frac{1}{4}$; W. $3\frac{1}{4}$; T. $2\frac{3}{4}$. Southern, N. to Washington, D.C.

ORDER L.—GALLINÆ.

(The Gallinaceous Birds.)

Bill short, stout, convex, horny, not constricted; nostrils scaled or feathered, cutting edge of upper mandible overlapping. Head often partly or wholly naked, sometimes with fleshy processes. Legs moderate, stout; hind toe elevated (excepting in *Cracidæ*, etc.), smaller than the other toes, sometimes wanting. Tarsus broadly scutellate (sometimes feathered), occasionally spurred in the males; claws blunt, not much curved. Wings short, strong, concave; tail various, sometimes wanting, often immensely developed. Precocial, often polygamous.

A large order comprising the various kinds of domesticated fowl as well as the chief game birds of most countries.

FAMILY XLVIII.—MELEAGRIDÆ.

(The Turkeys.)

Large birds, with the head and neck unfeathered, covered with scattered hairs, and more or less carunculate. Bill moderate; nostrils bare; forehead with an elongate fleshy process. Tarsus spurred in male; hind toe elevated. Tail nearly as long as wing, truncate, of more than twelve feathers. Breast of male mostly with a tuft of long bristles. Genus one; species two. *M. ocellatus*, of tropical America, and the common Turkey.

1. *MELEAGRIS,* Linnæus. TURKEYS.

1. ***M. gallopavo,*** L. WILD TURKEY. Glossy, coppery black; L. 48; W. 21; T. 18½. Canada to Rocky Mountains, and south to Mexico, becoming extinct eastwards. The domestic Turkey is descended from a Mexican variety.

FAMILY XLIX.—TETRAONIDÆ.

(*The Grouse.*)

Nostrils and tarsus densely feathered. Toes usually naked. Tail various with sixteen to twenty feathers. Usually a naked strip over eye; sides of neck often with a bare patch or lengthened feathers, or both. Genera seven; species fifteen; chiefly North American.

* Toes naked.

† Tarsus feathered to the toes.

‡ Tail of 16 feathers; colors dark. . . . CANACE, 1.

‡‡ Tail of 18 feathers.

a. Neck without peculiar feathers; middle tail feathers projecting. . . . . PEDIŒCETES, 2.

aa. Neck with peculiar, elongated, lanceolate feathers above a large, bare, bright-colored patch. . CUPIDONIA, 3.

†† Tarsus feathered about half way; tail of 18 soft, broad feathers. . . . . BONASA, 4.

** Toes feathered; winter plumage pure white. . LAGOPUS, 5.

1. ***CANACE,*** Reichenbach. AMERICAN GROUSE.

< *Tetrao,* L.

1. ***C. canadensis,*** (L.) Reich. SPRUCE PARTRIDGE. CANADA GROUSE. Black above with plumbeous markings; mostly black below with white spots; tail with an orange brown terminal band; ♀ smaller, black interrupted or streaky; L. 16; W. 6¾; T. 5½. Spruce swamps, N. U. S. and Northward.

2. PEDIŒCETES, Baird. SHARP-TAILED GROUSE.

1. ***P. phasianellus,*** (L.) var. ***columbianus,*** (Ord.) Coues. SHARP-TAILED GROUSE. Chiefly yellowish brown and white; sexes alike; L. 18; W. 8¾; T. 5. Illinois to Colorado, N. and W.

3. *CUPIDONIA,* Reich. Pinnated Grouse.

1. ***C. cupido,*** (L.) Baird. Prairie Hen. Prairie Chicken. Sides of neck with a tuft of long pointed feathers, beneath which is a patch of bare, red skin, capable of great inflation; black, tawny and white, barred and streaked; L. 17; W. 9; T. 4½; ♀ smaller. Prairies, etc., Martha's Vineyard to La. and N.; nearly exterminated eastward.

4. *BONASA,* Stephens. Ruffed Grouse.

1. ***B. umbellus,*** (L.) Stephens. Partridge (North.) Pheasant (South.) Crested; sides of neck with a ruff of soft black feathers; variegated, reddish or grayish brown, with blackish and pale; L. 18; W. 7¼; T. 7. E. U. S., abundant in woodland.

5. *LAGOPUS,* Vieillot. Ptarmigans.

1. ***L. albus,*** (Gm.) Aud. White Ptarmigan. Willow Grouse. Fore parts cinnamon brown, variegated with blackish; in winter pure white; bill stout; L. 16; W. 8; T. 5. British America; N. U. S. (rarely, in winter.)

FAMILY L.—PERDICIDÆ.

(The Partridges.)

Nostrils unfeathered, protected by a naked scale; tarsus bare and scutellate, circumorbital space usually not bare; in most respects similar to the Grouse, but smaller. Our species are crested (excepting the common Quail) and Western or Southwestern.

1. *ORTYX,* Stephens. Bob-Whites.

1. ***O. virginianus,*** (L.) Bon. Quail (North.) Partridge (South.) Bob-White. Forehead, line through eyes, chin and throat white, brownish yellow in ♀;

crown dark; plumage generally chestnut red, barred and streaked; L. 9½; W. 5; T. 3. E. U. S., and West Indies; W. to Plains.

ORDER M.—LIMICOLÆ.

(*The Shore Birds.*)

Tibia more or less naked below (sometimes very slightly); legs, and usually neck also, elongated; hind toe free and elevated, often wanting. Head globose, abruptly sloping to the base of the bill, completely feathered (except in *Philomachus* ♂); gape short; bill weak, flexible, more or less soft-skinned, and therefore sensitive, blunt at tip, without hard cutting edges—fitted for probing in the mud; nostrils slit-like, surrounded by soft skin, never feathered; body never strongly compressed or depressed; nature precocial.

Birds of medium or small size, more or less aquatic; found in most regions; very abundant in America.

FAMILY LI.—CHARADRIIDÆ.

(*The Plovers.*)

Head rather large, nearly globose; bill of moderate length, shaped somewhat like a pigeon's bill, with a constriction behind the horny terminal portion; nasal fossæ lined with soft skin, through which the slit-like nostrils open. Wings long and pointed, usually reaching beyond the tip of the short tail, sometimes spurred. Toes usually three, with basal web; tarsus reticulated; tibiæ naked below. Sexes similar, but seasonal changes of plumage great. Species sixty or more, in most parts of the world.

* Plumage speckled; black below in breeding season.

† Hind toe present, very short. . . . SQUATAROLA, 1.

†† Hind toe absent. . . . . . . CHARADRIUS, 2.

** Plumage not speckled; head and neck with dark bands in the breeding season; toes 3. . . . . ÆGIALITIS, 3.

1. SQUATAROLA, Cuvier. WHISTLING PLOVERS.

1. ***S. helvetica,*** (L.) Cuv. BLACK-BELLIED PLOVER. OX-EYE. Grayish, speckled; black below in breeding season, at other times white; L. 11½; W. 7; T. 3; B. 1¼; Ts. 2. In most parts of the world.

2. CHARADRIUS, L. GOLDEN PLOVERS.

1. ***C. fulvus*** (Gm.) var. ***virginicus,*** (Borck.) Coues. GOLDEN PLOVER. FROST BIRD. Dark and grayish above, profusely speckled, some of the spots bright yellow; black below in breeding season, at other times grayish; L. 10½; W. 7; T. 3; B. 1; Ts. 1⅝. N. Am., a well known game bird.

3. ÆGIALITIS, Boie. RING-NECK PLOVERS.

* Bill black, rather long; L. 8 or more.

1. ***Æ. vociferus,*** (L.) Cass. KILDEER PLOVER. Brown; rump bright orange brown; tail with black, white, and orange; two black bars across breast, and one above the white forehead; L. 9½; W. 6¼; T. 3½. N. Am., abundant in the Miss. Valley.

** Bill black-tipped, short and stout; L. 7 or less.

2. ***Æ. semipalmatus,*** (Bon.) Cab. RING-NECK PLOVER. Dark ashy brown; black bands broad; feet semipalmate; L. 7. N. Am.

3. ***Æ. melodus,*** (Ord.) Cab. PIPING PLOVER. Very pale ashy brown, clear white below; dark bands narrow and faint; toes slightly webbed; L. 6¾. E. N. Am., abundant along the coast.

FAMILY LII.—HÆMATOPODIDÆ.

(*The Turnstones.*)

Bill hard, acute, or truncate; nasal fossæ short, broad, and shallow. Legs short, stout, brightly colored. Genera two, not much alike; species six or eight; in most parts of the world.

* Toes 3, webbed at base; tarsus reticulate, shorter than the truncate, compressed, almost woodpecker-like bill. HÆMATOPUS, 1.

** Toes 4, not webbed; tarsus scutellate in front, as long as the sharp, pointed bill. . . . . . . STREPSILAS, 2.

1. ***HÆMATOPUS,*** Linnæus. OYSTER CATCHERS.

1. ***H. palliatus,*** Temminck. OYSTER CATCHER. Ashy brown and blackish, mostly white below; L. 18; W. 10; T. 4½; B. 3. Coasts.

2. ***STREPSILAS,*** Linnæus. TURNSTONES.

1. ***S. interpres,*** (L.) Illiger. TURNSTONE. Variegated; black, white, brown, and chestnut above; mostly white below; no reddish in winter; L. 8½; W. 6; T. 2½. Cosmopolitan; abundant.

FAMILY LIII.—RECURVIROSTRIDÆ.

(*The Avocets.*)

Legs excessively long. Bill very slender, long, acute, often recurved. Genera three, species eight; in most parts of the world. *Himantopus* is said to have the longest legs relatively of any bird.

* Toes 4, full webbed; bill recurved, flattened, tapering to a needle-like point; plumage beneath thickened as in ducks; swimmers. . . . . RECURVIROSTRA, 1.

** Toes 3, semipalmate; bill nearly straight, not flattened. HIMANTOPUS, 2.

1. RECURVIROSTRA, Linnæus. Avocets.

1. ***R. americana,*** Gm. Avocet. Blue Stocking. White, marked with black and cinnamon; legs blue; L. 18; W. 8; T. 3½. U. S.

2. HIMANTOPUS, Brisson. Stilts.

1. ***H. nigricollis,*** Vieillot. Stilt. Long Shanks. Lawyer. Glossy black, white below, legs pink; L. 15; W. 9; T. 3; Ts. 4. U. S.

FAMILY LIV.—PHALAROPODIDÆ.

(The Phalaropes.)

Snipe-like birds with the toes lobed, as in the coots and grebes, but the lobes narrower. Swimmers; body depressed and the under plumage thick as in the Ducks. Tarsus much compressed. Three species representing as many genera; of northern regions of both hemispheres, southward in winter.

* Bill flattened; membranes scalloped. . . Phalaropus, 1.
** Bill subulate; membranes scalloped. . . Lobipes, 2.
*** Bill subulate; membranes plain. . . Steganopus, 3.

1. PHALAROPUS, Brisson. Red Phalaropes.

1. ***P. fulicarius,*** (L.) Bon. Red Phalarope. Variegated above, purplish chestnut below; young white below; L. 8; W. 5; T. 2¾; B. 1. Northern Am.

2. LOBIPES, Cuvier. Northern Phalaropes.

1. ***L. hyperboreus,*** (L.) Cuv. Northern Phalarope. Grayish black, variegated; rump and under parts white; sides of neck with chestnut stripe. Northern regions.

3. STEGANOPUS, Vieillot. Phalaropes.

1. ***S. wilsoni,*** (Sab.) Coues. Wilson's Phalarope.

Ashy above, variegated; rump and under parts white; sides of neck with a black stripe which changes to chestnut below. Northern regions.

FAMILY LV.—SCOLOPACIDÆ.

(*The Snipe.*)

Bill elongated, usually longer than the head; if short not plover-like, being soft-skinned throughout (hard when dry); nasal grooves narrow channels ranging from half to nearly the whole length of the bill; sides of lower mandible usually also grooved; nostrils narrow exposed slits; head feathered. Wings usually thin and pointed; tail short and soft; tibiæ rarely entirely feathered. Tarsus never entirely reticulate and usually scutellate in front and behind; hind toe present (except in *Calidris*); front toes cleft or slightly webbed; size medium or small. Sexes alike or female slightly larger; seasonal changes in plumage often strongly marked. Eggs usually four, placed with the small ends together in a slight nest or depression in the ground; notes various; mostly migratory or gregarious. Genera fifteen or more, species about ninety; chiefly of northern regions, but some species in most parts of the world.

* Toes 3. . . . . Calidris, 8.

** Toes 4.

† Tarsus scutellate in front only; bill slender, decurved, very much longer than the head. . . . Numenius, 15.

†† Tarsus scutellate in front and behind.

‡ Feet semipalmate; toes somewhat webbed at base.

a Tail barred crosswise, with light and dark colors.

b. Gape not reaching beyond base of culmen.

c. Culmen furrowed; length less than 12. Macrorhamphus, 4.

cc. Culmen unfurrowed; length more than 12. LIMOSA, 9.

bb. Gape reaching beyond base of culmen.

d. Length more than 9.

e. Bill longer than head. . . . TOTANUS, 10.

ee. Bill not longer than head.

f. Tail more than half the length of wing. ACTITURUS, 13.

ff. Tail less than half the length of wing. PHILOMACHUS, 12.

dd. Length less than 9; second toe unwebbed.

g. Bill grooved nearly to tip; back not speckled with white; adult with black spots below. TRINGOIDES, 11.

gg. Bill grooved about half way to tip; back speckled with white, not spotted below. . TOTANUS, 10.

aa. Tail not barred.

h. One minute web; primaries mottled with black. TRYNGITES, 14.

hh. Feet with two plain webs.

i. Bill about as long as head. . . EREUNETES, 6.

ii. Bill much longer than head. . MICROPALAMA, 5.

‡‡ Toes not webbed at all.

j. First primary attenuate; bill straight, shorter than head; culmen grooved. . . . . . SCOLOPAX, 2.

jj. First three primaries attenuate; bill as in *Scolopax*. PHILOHELA, I.

jjj. Primaries not attenuate.

k. Bill straight, about twice as long as head. GALLINAGO, 3.

kk. Bill straight, much shorter than head; primaries mottled with black. . . . TRYNGITES, 14.

kkk. All other Sandpipers. . . . . TRINGA, 7.

1. *PHILOHELA*, Gray. AMERICAN WOODCOCKS.

1. ***P. minor***, (Gm.) Gray. AMERICAN WOODCOCK.

Variegated, black, brown, gray, and russet; below warm brown; eye high and far back; L. 11; W. 5; B. 3; T. 1¼. E. U. S., in swamps, etc.

2. *SCOLOPAX*, Linnæus. EUROPEAN WOODCOCKS.

1. ***S. rusticola***, L. EUROPEAN WOODCOCK. General appearance of *Philohela*, but a third larger. European; accidental on our Atlantic coast.

3. *GALLINAGO*, Leach. SNIPE.

1. ***G. wilsoni***, (Temm.) Bon. AMERICAN SNIPE. WILSON'S SNIPE. Back varied with black and bay; crown black, with a pale median stripe; bill straight, very long; L. 11; W. 5; B. 2½; leg naked, 3; T. 2⅓. E. U. S., abundant.

4. *MACRORHAMPHUS*, Leach. RED-BREASTED SNIPE.

1. ***M. griseus***, (Gm.) Leach. GRAY SNIPE. BROWN-BACK. Blackish and grayish; breast bay in summer; bill long nearly as in *Gallinago;* L. 11; W. 5½; T. 2½. North America; abundant coastwise.

5. *MICROPALAMA*, Baird. STILT SANDPIPERS.

1. ***M. himantopus***, (Bon.) Baird. STILT SANDPIPER. Blackish, marked with chestnut, etc.; ashy gray in winter; bill nearly as in *Gallinago;* L. 9; W. 5; T. 2¼. N. Am., not abundant.

6. *EREUNETES*, Illiger. SAND-PEEPS.

1. ***E. pusillus***, (L.) Cass. SEMIPALMATED SANDPIPER. PEEP. Plumage various, usually pale, white below; small; L. 6½; W. 3¾; T. 2. N. Am.; abundant along beaches.

7. TRINGA, Linnæus. SANDPIPERS.

* Bill, tarsus, and middle toe with claw, of about equal length. (*Actodromas.*)

† Upper tail coverts (except the lateral series) black or dark brown; throat with an ashy or brownish suffusion and dusky streaks.

1. ***T. minutilla,*** Vieill. LEAST SANDPIPER. PEEP. Smallest of the Sandpipers, resembling *Ereunetes*, but the feet different; L. 6; W. 3½; T. 2. N. Am., abundant.

2. ***T. bairdii,*** (Coues) Scl. BAIRD'S SANDPIPER. Colors of preceding but larger; throat but little streaked; L. 7 to 7½; W. 4⅝; T. 2¼; B. ⅞. America, rare E. of the Mississippi R.

3. ***T. maculata,*** Vieill. PECTORAL SNIPE. JACK SNIPE. GRASS SNIPE. Crown unlike neck; throat ashy-shaded and sharply streaked; L. 9; W. 5½; B. 1⅛. N. Am., abundant.

†† Upper tail coverts white, with or without dusky marks; throat sharply streaked, with little if any ashy suffusion.

4. ***T. fuscicollis,*** Vieill. WHITE-RUMPED SANDPIPER. L. 7½; W. 4¾; T. 2¼. E. U. S., abundant along the coast.

** Bill, tarsus, and middle toe, obviously not of equal length.

‡ Tarsus shorter than middle toe; tibiæ feathered. (*Arquatella.*)

5. ***T. maritima,*** Brünnich. PURPLE SANDPIPER. Ashy black with purplish reflections; feathers with pale edgings; lower parts, etc., mostly white; bill nearly straight; L. 9; W. 5; T. 2⅔; B. 1¼. Atlantic Coast.

‡‡ Tarsus not shorter than middle toe; tibiæ bare below.

a. Bill slightly decurved, much longer than tarsus. (*Pelidna.*)

6. ***T. alpina*** (L.) var. ***americana,*** Cass. AM. DUNLIN. OX-BIRD. BLACK-BELLIED SANDPIPER. Chestnut brown above; feathers black centrally; belly, in summer,

with a broad black area; L. 9; W. 5; T. $2\frac{1}{3}$; B. $1\frac{3}{4}$. N. Am.

aa. Bill perfectly straight. (*Tringa.*)

7. ***T. canutus,*** L. ROBIN SNIPE. RED-BREASTED SANDPIPER. Brownish black, brownish red (robin-like) below; L. 11; W. $6\frac{1}{2}$; T. $2\frac{1}{3}$. Atlantic Coast; abundant.

8. *CALIDRIS,* Cuvier. SANDERLINGS.

1. ***C. arenaria,*** (L.) Ill. SANDERLING. RUDDY PLOVER. Variegated; form of *T. canutus*, but the hind toe wanting; L. 8; W. 5; T. $2\frac{1}{4}$; B. 1. N. Am.; abundant coastwise.

9. *LIMOSA,* Brisson. GODWITS.

1. ***L. fedoa,*** (L.) Ord. GREAT MARBLED GODWIT. MARLIN. Cinnamon brown, variegated above, nearly uniform below; tail barred; no pure white; L. 16 to 22; W. 9; T. $3\frac{1}{2}$; B. $4\frac{1}{2}$. U. S., abundant along shores.

2. ***L. hudsonica,*** (Lath.) Sw. BLACK-TAILED GODWIT. Brownish black and reddish, more or less variegated above and below; some white; tail black, white at base; L. 15; W. 8; Ts. $2\frac{1}{2}$; B. $3\frac{1}{2}$. N. Am., rather northerly.

10. *TOTANUS,* Bechstein. TATTLERS.

* Toes with two sub-equal webs; legs dark or bluish. (*Symphemia.*)

1. ***T. semipalmatus,*** Gmelin. WILLET. SEMIPALMATED TATTLER. Grayish, variegated; L. 12 to 16; W. $7\frac{1}{2}$; T. 3; B. $2\frac{1}{2}$. U. S., common coastwise.

** Toes with the inner web very small; legs yellow. (*Glottis*, Nilsson.)

2. ***T. melanoleucus,*** Gm. GREATER TELL-TALE. YELLOW SHANKS. STONE SNIPE. Ashy brown, variegated; bill very slender; legs long; L. $12\frac{1}{2}$; W. $7\frac{1}{2}$; T. $3\frac{1}{4}$; B. $2\frac{1}{4}$. N. Am., frequent.

3. ***T. flavipes,*** Gm. LESSER TELL-TALE. YELLOW SHANKS. Colors as in preceding; smaller; legs longer; L. 11; W. 6½; T. 2½; B. 1¾. U. S., abundant.

*** Toes with inner web rudimentary; legs blackish. (*Rhyacophilus*, Kaup.)

4. ***T. solitarius,*** Wilson. SOLITARY TATTLER. Olive brown, streaked and speckled with whitish above; below white, breast with dusky suffusion; bill straight and slender; L. 9; W. 5; T. 2½; B. 1¼. U. S., abundant about secluded ponds, etc.

11. TRINGOIDES, Bonaparte. SPOTTED SANDPIPERS.

1. ***T. macularius,*** (L.) Gray. TIP-UP. TEETER-TAIL. SPOTTED SANDPIPER. Lustrous drab above, varied with black; pure white below, with round black spots in adult; L. 8; W. 4; T. 2; B. 1. U. S., every where.

12. PHILOMACHUS, Möhring. RUFFS.

1. ***P. pugnax,*** (L.) Gray. RUFF (♂). REEVE (♀). Male in breeding season with a great ruff, and the face bare; ♀ without these characters; L. 10; W. 7; T. 2¾; B. 1¼. European; accidental on our coasts.

13. ACTITURUS, Bonaparte. UPLAND SANDPIPERS.

1. ***A. bartramius,*** (Wilson) Bon. UPLAND PLOVER. Dark grayish, variegated; L. 13; W. 7; T. 4; B. 1¼. U. S., abundant in fields, etc.

14. TRYNGITES, Cabanis. BUFF-BREASTED SANDPIPERS.

1. ***T. rufescens,*** (Vieill.) Cab. Grayish, reddish below; quills with white and finely mottled with black; L. 8; W. 5½; T. 2¼. U. S., with the last, but not common.

15. NUMENIUS, Linnæus. CURLEWS.

1. ***N. longirostris,*** Wils. LONG-BILLED CURLEW.

SICKLE BILL. Reddish gray, variegated; L. 24; W. 12; T. 4; B. 5 to 9. U. S., frequent.

2. ***N. hudsonicus,*** Lath. JACK CURLEW. Similar, but paler; L. 18; W. 9; T. 3½; B. 3 or 4. U. S., and northward.

3. ***N. borealis,*** (Forst.) Lath. ESQUIMAUX CURLEW. DOUGH BIRD. More reddish; L. 15 or less; W. 8½; T. 3; B. 2½. U. S., northwards.

ORDER N.—HERODIONES.

(*The Herons and Storks.*)

Birds usually of large stature, with compressed body, long legs and a very long "S-bent" neck; tibia naked below; toes long and slender, cleft or slightly webbed, the hind toe comparatively long and (usually) not elevated, provided with a large claw. Wings broad, rounded. Tail short. Head narrow, gradually contracting to the stout base of the bill, which is long and mostly hard and acute, with sharp cutting edges; lores, orbital space, and often whole head naked. Plumage with powder-down tracts (explained below); altricial.

FAMILY LVI.—ARDEIDÆ.

(*The Herons.*)

Large birds with the bill straight, longer than the head, compressed, acute, with sharp-cutting edges; upper mandible grooved; nostrils linear; lores naked, the bill appearing to run directly to the eyes; rest of head feathered; parts of the body with "powder-down tracts,"—strips of short, dusty, or greasy down-like feathers, usually three pairs of these strips, *i. e.*, on the back above the hips, on the belly under the hips, and on the breast; usually long plumes from the back or head in the breeding season. Wings broad. Tail very short. Tibiæ largely naked below; toes long and slender, hind toe on a level with the rest, middle claw pectinate. Sexes usually colored alike. Species nearly one hundred; in most parts of the world, abundant in the warmer regions.

* Tail of 12 feathers; usually a crest or train in the breeding season; lateral toes more than half length of tarsus.

† Tibia bare two inches or more.

‡ Length 36 or more.

a. General color bluish or ashy brown. . ARDEA, 1.

aa. Color white at all times. . . . HERODIAS, 2.

‡‡ Length 24 or less.

b. Color white at all times; legs black and yellow. GARZETTA, 3.

bb. General color bluish (young white), legs black or bluish. FLORIDA, 4.

†† Tibia bare one inch or less.

c. Tarsus shorter than middle toe and claw.

d. Bill more than thrice as long as high. . BUTORIDES, 5.

dd. Bill not three times as long as high. NYCTIARDEA, 6.

cc. Tarsus longer than middle toe and claw; bill more than half an inch deep at base. . . NYCTHERODIUS, 7.

** Tail of 10 feathers; no crest nor train; lower neck bare behind; length less than 30.

e. Length more than 18; tawny, much streaked. BOTAURUS, 8.

ee. Length less than 18; glossy blackish or chestnut. ARDETTA, 9.

1. ARDEA, Linnæus. HERONS.

1. ***A. herodias***, L. GREAT BLUE HERON. Grayish blue, marked with black and white; back of head crested in breeding season; tibia and edge of wing chestnut brown; L. 48; W. 20; T. 7; B. 5½; Ts. 6½; ♀ much smaller. U. S., common.

2. *HERODIAS*, Gray. Great White Egrets.

1. ***H. egretta*,** (Gm.) Gray. Great White Egret. White Heron. Pure white; head without lengthened feathers; back in breeding season with a long train; L. 40; W. 17; B. 5; Ts. 6. U. S., chiefly southerly.

3. *GARZETTA*, Bonaparte. Little White Egrets.

1. ***G. candidissima*,** (Jacq.) Bon. Snowy Egret. Pure white; head and neck with long plumes in breeding season; L. 24; W. 12; B. 3; T. 4. Southern and middle States; abundant.

4. *FLORIDA*, Baird. Little Blue Herons.

1. ***F. cœrulea*,** (L.) Baird. Little Blue Heron. Slaty blue; young white; head with elongated feathers; no dorsal plumes; L. 24; W. 12; B. 3; Ts. 4. U. S.; abundant, southerly.

5. *BUTORIDES*, Bonaparte. Green Herons.

1. ***B. virescens*,** (L.) Bon. Green Heron. Crown, back and wings lustrous dark green; neck purplish cinnamon; crested; back with lengthened feathers; L. 18; W. 7; B. 2½. U. S., abundant.

6. *NYCTIARDEA*, Swainson. Night Herons.

1. ***N. grisea*** (L.) Steph., var. ***nœvia*,** (Bodd.) Allen. Qua Bird. Squawk. Night Heron. Bluish gray, crown and shoulders glossy green; no peculiar feathers save two or three long, white occipital plumes; young speckled, very different; L. 24; W. 14; B. 3; Ts. 3. U. S., frequent.

7. *NYCTHERODIUS*, Auctorum. Yellow-Crowned Night Herons.

1. ***N. violaceus*,** (L.) ——. Yellow-Crowned Night

HERON. Grayish plumbeous; crested; back with long plumes; crown, etc., tawny or white; young speckled; size of last; B. 2¾; Ts. 3¾. U. S.

8. BOTAURUS, Stephens. BITTERNS.

1. ***B. mugitans,*** (Bartr.) Coues. INDIAN HEN. STAKE DRIVER. BITTERN. Tawny brown of various shades, excessively variegated every where; dark patch on each side of neck; L. 23 to 28; W. 12; T. 4½; B. 3. U. S., abundant.

9. ARDETTA, Gray. LEAST BITTERN.

1. ***A. exilis,*** (Gm.) Gray. LEAST BITTERN. ♂ chiefly glossy greenish black above, brownish yellow below, neck and shoulders with chestnut; ♀ with purplish chestnut instead of black; L. 14; W. 5; T. 1¾; B. 1¾. U. S., rather rare.

FAMILY LVII.—TANTALIDÆ.

(The Ibises.)

Stork-like birds, usually of large size, with the head more or less bare of feathers when adult; neck and legs long; body small. Wings large and rounded. Tail very short. Tibia bare for some distance; toes 4; hind toe lengthened and low down. Genera four; species fifteen; swamps and lakes of warm regions. Sexes alike. Allied to the Storks (*Ciconiidæ*) of the Old World.

* Tarsus reticulate; bill very stout, tapering, decurved. TANTALUS, 1.

** Tarsus scutellate in front; bill grooved, curved (curlew-like.) IBIS, 2.

1. TANTALUS, Linnæus. WOOD IBISES.

1. ***T. loculator,*** L. WOOD IBIS. White; quills, tail and primary coverts black; bare part of head and neck

bluish; L. 48; W. 20; B. 9. Southern States, N. to Ohio and Colorado.

2. *IBIS,* Möhring. IBISES.

* Claws curved. (*Ibis.*)

1. ***I. alba,*** (L.) V. WHITE IBIS. Pure white, wings with black; L. 24; W. 11; T. 4; B. 7. Southern States, N. to L. I.

** Claws nearly straight. (*Falcinellus.*)

2. ***I. falcinellus*** var. ***ordii,*** (Bon.) Allen. GLOSSY IBIS. Rich dark chestnut, with greenish and purplish on head; L. 24; W. 11; T. 4; B. 4½. S. States, N. to N. England.

ORDER O.—ALECTORIDES.

(*The Cranes and Rails.*)

Tibia naked below; neck, legs and feet much as in *Herodiones*, except that the hind toe is small and elevated, and provided with a small claw. Bill various, usually lengthened; head fully feathered or else extensively bald. Body more or less compressed. Wings short, rounded, concave. Tail very short and small; size various.

FAMILY LVIII.—GRUIDÆ.

(*The Cranes.*)

Very large birds with the head and neck extremely long. Wings large. Tail short. Head more or less naked, with scattered hair-like feathers. Plumage mostly compact. Bill as long or longer than head, straight and slender; tibiæ extensively naked; tarsus scutellate; toes rather short; hind toe highly elevated. Genera three; species fourteen, of various parts of the world.

1. GRUS, Linnæus. Cranes.

1. ***G. americanus,*** (L.) Ord. White or Whooping Crane. Adult pure white with black on wings; bare part of head very hairy; young grayish, the head feathered; L. 50; W. 24; T. 9; Ts. 12; B. 6. U. S., rather southerly.

2. ***G. canadensis,*** (L.) Temm. Brown or Sand-Hill Crane. Plumbeous gray, never whitening; head sparsely hairy; smaller. U. S., chiefly S. and W.

FAMILY LIX.—RALLIDÆ.

(*The Rails.*)

Birds of medium or small size, with compressed bodies and muscular legs. Wings and tail short. Hind toe short and elevated; front toes very long. Bill various, rather short. Plumage blended. Sexes alike. Species about one hundred and fifty, of most parts of the world.

* Forehead feathered; no frontal plate. (Rallinæ.)

† Bill decurved, longer than head. . . . Rallus, 1.

†† Bill straight, shorter than head. . . . Porzana, 2.

** Forehead covered with a broad, horny, frontal plate.

‡ Toes scarcely or not lobate. (Gallinulinæ.)

a. Nostrils linear; tarsus less than 2. . Gallinula, 3.

aa. Nostrils nearly circular; tarsus about 2. Porphyrio, 2.

‡‡ Toes lobate, edged with broad flaps. (Fulicinæ.) Fulica, 5.

1. RALLUS, Linnæus. Rails.

1. ***R. longirostris,*** Bodd. Clapper Rail. Salt-Water Marsh Hen. Olive brown, variegated with ashy; dull reddish brown below; L. 14 to 16; W. 6; T. $2\frac{1}{4}$; B. $2\frac{1}{2}$; ♀ smaller. Salt marshes; rather southerly.

2. ***R. elegans,*** Aud. King Rail. Fresh-Water Marsh Hen. Brownish black, with chestnut below

and on wing coverts; much brighter colored than the last, and rather larger. U. S., fresh-water marshes.

3. ***R. virginianus,*** L. VIRGINIA RAIL. Colors exactly as in *R. elegans;* much smaller; L. 10; W. 4; T. 1½; B. 1½. U. S., frequent.

2. *PORZANA,* Vieillot. LITTLE RAILS.

1. ***P. carolina,*** (L.) V. CAROLINA RAIL. SORA. "ORTOLAN." Olive-brown, variegated; face and middle line of throat black; breast slaty gray; back streaked; belly barred; L. 9; W. 4½; T. 2. U. S., not rare.

2. ***P. noveboracensis,*** (Gm.) Cass. YELLOW RAIL. Variegated above; L. 6; W. 3¼; T. 1½. E. U. S., not common.

3. ***P. jamaicensis,*** (Gm.) Cass. BLACK RAIL. Blackish; L. 5½. S. Am., etc., rarely in U. S.

3. *GALLINULA,* Brisson. GALLINULES.

1. ***G. galeata,*** (Licht.) Bon. FLORIDA GALLINULE. Brownish olive above, grayish black on head and below; bill, frontal plate and ring around tibia red; L. 15; W. 7½; T. 3½; Ts. 2. S. States, straying northward. (Manitowoc, Wis. *Jordan.*)

4. *PORPHYRIO,* Temminck. PURPLE GALLINULES.

1. ***P. martinica,*** (L.) Temm. PURPLE GALLINULES. Olive green; head and below purplish blue; crissum white; bill mostly red; L. 12; W. 7; T. 3. S. States, N. to Maine.

5. *FULICA,* Linnæus. COOTS.

1. ***F. americana,*** Gm. COOT. MUD HEN. Dark slate color or sooty; bill brownish; L. 14; W. 8; T. 2. U. S., abundant in reedy swamps; swims well.

ORDER P.—LAMELLIROSTRES.

(*The Anserine Birds.*)

Bill lamellate, *i. e.*, furnished along each cutting edge with a regular series of tooth-like processes, which correspond to certain laciniate processes of the fleshy tongue, which ends in a horny tip; bill large, thick, high at base, depressed towards the end, membranous except at the obtuse tip which is occupied by a horny nail; no gular pouch. Head high, compressed, with sloping forehead; eyes small. Feet 4-toed (excepting some Flamingoes), palmate; hind toe small, elevated; tibia feathered in *Anatidæ*, bare below in the Flamingoes. Wings strong, short. Legs short (except in *Phœnicopteridæ*, where excessively elongated); precocial; swimming birds. An important and familiar order, comprising nearly all the "Water Fowl" which are valued in domestication or as game birds. There are two families, *Phœnicopteridæ* the Flamingoes, and the following:

FAMILY LX.—ANATIDÆ

(*The Ducks.*)

Characters of bill, etc., as given above. Body heavy, flattened beneath. Head large; eyes small. Tail various, usually short, of 14 to 16 feathers, the lower coverts being long and full. Feet short, anterior toes full-webbed. Sexes usually quite unlike (excepting among the Swans and Geese.) Species one hundred and seventy-five, of all parts of the world; migratory. The sub-families are indicated below.

* Lores naked; adult entirely white; large birds; Swans. (CYGNINÆ.) . . . . CYGNUS, 1.

** Lores feathered; tarsus entirely reticulate; Geese. (ANSERINÆ.)

a. Bill and legs not black; colors white, bluish, etc. ANSER, 2.

aa. Bill and legs black; neck black. . . BRANTA, 3.

*** Lores feathered; tarsus scutellate in front; Ducks.

† Bill depressed; the lamellæ simple, bluntish.

‡ Hind toe simple, not bordered by membrane. "River Ducks." (ANATINÆ.)

b. Head crested; tip of bill formed entirely by the nail; colors brilliant in ♂. . . . . . . AIX, 11.

bb. Bill very much widened towards the tip; speculum green. . . . . SPATULA, 10.

bbb. Head not crested; bill not much widened towards tip.

c. Tail wedge-shaped, at least ½ length of wing. ♂ of DAFILA, 5.

cc. Tail less than half length of wing.

d. Bill shorter than head.

e. Crown streaked; tail feathers acute. DAFILA, 5.

ee. Crown creamy or white; speculum green. MARECA, 7.

dd. Bill about as long as head; speculum white; wing coverts chestnut. . . CHAULELASMUS, 6.

ddd. Bill a little longer than head.

f. Speculum violet, bordered with black and white. ANAS, 4.

ff. Speculum green.

g. Wing coverts sky blue; head of ♂ plumbeous or purplish. . . . QUERQUEDULA, 8.

gg. Wing without blue; head of ♂ chestnut, with green band; almost crested. . NETTION, 9.

‡‡ Hind toe lobed (bordered by membrane.) "Sea Ducks." (FULIGULINÆ.)

h. Cheeks bristly; colors black and white (or gray.) CAMPTOLÆMUS, 16.

hh. Tail pointed, longer than wings (in adult); bill black and orange. . . . . . HARELDA, 15.

hhh. Tail rounded; the feathers stiff, narrow, exposed nearly to their bases, the upper coverts being very short. ERISMATURA, 20.

hhhh. Ducks with none of the above peculiarities.

i. Upper mandible gibbous at its unfeathered base; black or brown. . . . . . . ŒDEMIA, 19.

ii. Upper mandible not gibbous where unfeathered.

j. Nail at tip of bill narrow and distinct.

k. Head black or brown; nostrils sub-basal; bill longer than tarsus. . . . FULIX, 12.

kk. Head reddish or brownish, without white; nostrils nearly median; bill longer than tarsus. AYTHYA, 13.

kkk. Head black or gray, with white; nostrils nearly median; bill about as long as tarsus. BUCEPHALA, 14.

jj. Nail broad, scarcely distinct.

l. Feathers extending on culmen and partly on sides of upper mandible. . . SOMATERIA, 18.

ll. Feathers not extending on culmen; bill small, much tapering. . . HISTRIONICUS, 17.

†† Bill narrow, nearly cylindrical; the lamellæ acute, recurved, like saw-teeth; usually crested. Fish Ducks. (MERGINÆ.)

m. Bill not black; tarsus more than half the length of middle toe. . . . . MERGUS, 21.

mm. Bill black; tarsus half length of middle toe. LOPHODYTES, 22.

1. ***CYGNUS,*** Linnæus. SWANS.

> *Olor*, Wagler.

1. ***C. buccinator,*** Rich. TRUMPETER SWAN. Tail (normally) 24 feathered; bill without yellow spot, longer than head; nostrils sub-basal; L. about 50. Miss. Valley, W. and N.

2. ***C. columbianus,*** (Ord.) Coues. WHISTLING SWAN. Tail 20 feathered; bill with a yellow spot, not longer than head; nostrils median; L. 50. N. Am. (*C. americanus*, Sharpless.)

2. *ANSER,* Linnæus. Geese.

1. ***A. albifrons,*** Gm., var. ***gambeli,*** (Hartl.) Coues. White-Fronted Goose. Speckle-Bill. White or gray, blotched with black; back dark; head and neck grayish brown; forehead white in adult; claws pale; lamellæ usual; L. 27; W. 17; T. 6; Ts. 3. N. Am.

2. ***A. cœrulescens,*** L. Blue Goose. Size and form of next, but plumage ashy, varied with dark brown. N. Am., rather rare.

3. ***A. hyperboreus,*** Pallas. Snow Goose. Adult pure white or washed with reddish; wings with black; claws dark; young bluish; lamellæ very prominent; L. 30; W. 19; T. 6½; B. 2½.

3. *BRANTA,* Scopoli. Brant Geese.

= *Bernicla*, most authors.

1. ***B. bernicla,*** (L.) Brant Goose. Head, neck, front, quills, and tail, black; white patch on neck; white on rump, crissum, etc.; back brownish gray; L. 24; W. 13; T. 5; B. 1⅛. Northern States; in winter to Carolina.

2. ***B. canadensis,*** (L.) Wild Goose. Canada Goose. Grayish brown, paler below; head and neck black; white throat patch, extending on sides of head; tail black; upper coverts white; L. 36; W. 20; T. 7½; B. 2. N. Am., abundant; U. S. in winter.

4. *ANAS,* Linnæus. Ducks.

1. ***A. boschas,*** L. Mallard Duck. Tame Duck. ♂ head and upper neck rich glossy green, a white ring below; breast purplish chestnut; speculum violet; wing coverts tipped with black and white; ♀ duller, chiefly dull ochraceous, streaked with dark brown; L. 24; W. 12. Am., abundant; commonest westward. Original

of the common Domestic Duck; various hybrids of this species with others are described.

2. ***A. obscura,*** Gm. BLACK DUCK. Size of mallard and resembling the ♀, but darker; no decided white except under the wings. E. U. S., common.

5. *DAFILA,* Leach. PINTAIL DUCKS.

1. ***D. acuta,*** (L.) Jenyns. PIN-TAIL. SPRIG-TAIL. ♂ dark brown with purplish gloss; sides of neck with long white stripes; tail cuneate when developed, central feathers much projecting; ♀ speckled and streaked; tail shorter; L. 24; W. 11; T. 9 or less. N. Am.

6. *CHAULELASMUS,* Gray. GADWALLS.

1. ***C. streperus,*** (L.) Gray. GADWALL. ♂ barred, black and white, wing coverts chestnut, greater coverts black, speculum white; ♀ with similar markings; L. 22; W. 11. N. Am.

7. *MARECA,* Stephens. WIDGEONS.

1. ***M. penelope,*** (L.) Bon. EUROPEAN WIDGEON. Head and neck reddish brown; top of head brownish white; sides of head with green traces; L. 22; W. 11; T. 5. Europe; accidental in America, Wis. (*Kumlien*) and Atlantic Coast.

2. ***M. americana,*** (Gm.) Steph. AMERICAN WIDGEON. BALDPATE. Head and neck grayish, speckled; colors more emphatic; sides of head with bright green patch. N. Am., abundant.

8 *QUERQUEDULA,* Stephens. BLUE WINGED TEALS.

1. ***Q. discors,*** (L.) Steph. BLUE WINGED TEAL. ♂ head and neck blackish plumbeous, darkest on the crown; a white crescent in front of eye; under parts

thickly spotted; ♀ quite different, known by the wings; L. 16; W. 7; T. 3. E. U. S., to Rocky Mts.

9. *NETTION,* Kaup. Green-Winged Teals.

1. ***N. carolinensis,*** (Gm.) Kaup. Green-Winged Teal. A white crescent on sides in front of wings; shoulders plain; L. 15; W. 7½; T. 3½. N. America, common.

10. *SPATULA,* Boie. Shovellers.

1. ***S. clypeata,*** (L.) Boie. Shoveller. Spoon-Bill Duck. ♂ head and neck green; wing coverts blue; speculum green; ♀ with similar bill and wings; L. 20; W. 9½; B. 2¾. N. Am.

11. *AIX,* Swainson. Wood Ducks.

1. ***A. sponsa,*** (L.) Boie. Wood Duck. Summer Duck. Crested; ♂ head iridescent green and purple, with white stripes and a forked white throat patch; breast rich brownish; ♀ duller, head mostly gray; L. 20; W. 9½; T. 5. U. S. frequent; nesting in trees.

12. *FULIX,* Sundevall. Flocking-Fowl.

< *Fuligula*, Authors.

1. ***F. marila,*** (L.) Baird. Big Scaup Duck. Blue Bill. Raft Duck. Speculum white; no ring about neck; back and sides whitish, finely waved with black; ♀ face white; markings less distinct; L. 20; W. 9. N. Am.

2. ***F. affinis,*** (Eyton) Baird. Lesser Scaup Duck. Similar, but smaller; L. 16; W. 8. N. Am., rather southerly.

3. ***F. collaris,*** (Donovan) Baird. Ring-Necked Duck. Speculum gray; an orange brown collar about neck; ♀ without collar; L. 18; W. 8½. N. Am.

13. *AYTHYA,* Boie. CANVAS-BACK DUCKS.

1. ***A. ferina*** (L.) var. ***americana,*** (Eyton) Allen. RED HEAD. POCHARD. ♂ head and neck chestnut with red reflections; back mixed silvery and black; the dark waved lines unbroken; ♀ duller; bill shorter than head, two or less, bluish, the nail dark; L. 20; W. 10. N. Am., abundant.

2. ***A. vallisneria,*** (Wilson) Boie. CANVAS-BACK DUCK. Head more dusky; black wavy lines on back broken, the whitish predominating; bill as long as head, 2½ or more, dusky. N. Am.; especially coastwise in winter.

4. *BUCEPHALA,* Baird. GOLDEN-EYES.

Clangula, Authors.

1. ***B. clangula,*** (L.) Gray. GOLDEN-EYE. GARROT. ♂ head puffy, glossy green with some white; upper parts black; white continuous on outer surface of wing; ♀ head duller, snuff-colored and scarcely puffy; L. 16 to 19; W. 8 to 9. N. Am. and Europe.

2. ***B. islandica,*** (Gm.) Baird. BARROW'S GOLDEN-EYE. Similar; gloss of head purplish; white of wing divided by dark bar; more white on head; larger. N. U. S. and N.; rare.

3. ***B. albeola,*** (L.) Baird. DIPPER. BUFFLE-HEAD. BUTTER-BALL. SPIRIT DUCK. ♂ with head very puffy and iridescent; a large white ear patch; L. 16; W. 7; ♀ small, dark gray; head scarcely puffy. N. Am., abundant.

15. *HARELDA,* Leach. LONG-TAILED DUCKS.

1. ***H. glacialis,*** (L.) Leach. SOUTH-SOUTHERLY. OLD WIFE. LONG-TAILED DUCK. Reddish brown, nearly white in winter; tail very long; ♀ quite different, no

white on wing; L. 20; W. 9; T. 8, or less. N. Am. and Europe; chiefly northern and coastwise.

16. *CAMPTOLÆMUS,* Gray. Pied Ducks.

1. ***C. labradorius,*** (Gm.) Gray. Labrador Duck. ♂ chiefly black and white; ♀ plumbeous; L. 24; W. 9. Coast, chiefly northern; scarce.

17. *HISTRIONICUS,* Lesson. Harlequin Ducks.

1. ***H. torquatus,*** (L.) Bon. Harlequin Duck. ♂ leaden bluish, much varied; speculum violet and purple; ♀ dark brown, etc. Atlantic Coast, Am. and Europe.

18. *SOMATERIA,* Leach. Eider Ducks.

1. ***S. mollissima,*** (L.) Leach. Eider Duck. ♂ in breeding dress, white; under parts, rump, quills, and crown patch black; ♀ reddish brown, streaked; bill with long, club-shaped, frontal processes extending in line with culmen; L. 24; W. 12. Arctic regions; S. to New England in winter. (*S. dresseri*, Sharpe.)

2. ***S. spectabilis,*** (L.) Leach. King Eider. ♂ chiefly black; front parts, etc., white; frontal processes broad, squarish, out of line of culmen; slightly smaller. Northern regions; S. to N. Y.

19. *ŒDEMIA,* Fleming. Surf Ducks.

* Bill not encroached upon by frontal feathers; tail 16-feathered; no white on wings. (*Œdemia.*)

1. ***Œ. americana,*** Sw. American Black Scoter. ♂ entirely black; ♀ sooty brown, paler below and on throat; L. 18 to 24; W. 10. N. Am., all coasts.

** Bill broadly encroached upon by frontal feathers; a large white wing patch. (*Melanetta.*)

2. ***Œ. fusca,*** (L.) Sw. Velvet Scoter. White

WINGED SURF DUCK. ♂ black; white spot under eye; ♀ sooty brown, rather larger. Shores of Europe and N. Am.

*** Bill narrowly encroached upon by frontal feathers; no white on wings; tail 14-feathered. (*Pelionetta.*)

3. ***Œ. perspicillata,*** (L.) Fleming. SURF DUCK. SEA COOT. ♂ black, with white spot on forehead and nape; ♀ sooty brown; white patch on lores and cheeks; size of first. Coasts.

20. *ERISMATURA,* Bonaparte. STIFF TAILED DUCKS.

1. ***E. rubida,*** (Wils.) Bon. RUDDY DUCK. Chiefly brownish or tawny (reddish in perfect plumage), considerably waved and dotted; crissum white; L. 17; W. 6. N. Am., frequent.

2. ***E. dominica,*** (L.) Eyton. SAN DOMINGO DUCK. Smaller and redder; forehead and chin black; L. 13½; W. 6¼. S. America and W. Indies; accidental N. (Wis., *Kumlien*, L. Champlain, *Cabot.*)

21. *MERGUS,* Linnæus. MERGANSERS.

1. ***M. merganser,*** L. MERGANSER. GOOSANDER. FISH DUCK. ♂ black and white above, salmon-colored below; head glossy green, scarcely crested; ♀ smaller, ashy gray; head brownish; nostrils median; L. 24; W. 11. N. Am., common.

2. ***M. serrator,*** L. RED-BREASTED MERGANSER. FISH DUCK. Similar, but smaller and more crested; ♂ with throat reddish brown, black-streaked; wing with two black bars, instead of one as in last; nostrils sub-basal; L. 20; W. 9. N. Am., abundant.

22. *LOPHODYTES,* Reichenbach. CRESTED DIVERS.

1. ***L. cucullatus,*** (L.) Reich. HOODED MERGANSER.

Sheldrake. Black and white; sides chestnut in ♂; ♀ duller and grayish; crest high and compressed; nostrils sub-basal; L. 19; W. 8. N. Am., common.

ORDER Q.—STEGANOPODES.

(The Totipalmate Birds.)

Feet totipalmate; hind toe lengthened, scarcely elevated, united by a web to the other toes; tibiæ feathered Bill various, horny, never lamellate, cutting edges often serrate; nostrils very small or abortive; a prominent, naked, gular pouch; tarsus reticulate; altricial.

We here omit the families *Sulidæ* (Gannets) and *Tachypetidæ* (Frigate Birds) as they are exclusively marine. One species of the Southern family *Plotidæ*, (the Darters) *Plotus anhinga*, L., the Snake Bird or Water Turkey of the Southern swamps, sometimes comes up the Miss. R. to S. Ills.

FAMILY LXI.—PELECANIDÆ.

(The Pelicans.)

Large fish-eating birds, with very long bills which end in a claw-like hook; the broad space between the branches of the lower jaw occupied by a huge membranous sack; nostrils abortive; wings very long; tail very short; gregarious; sexes alike. Genus one; species six; found in most warm regions.

1. **PELECANUS,** Linnæus. Pelicans.

1. ***P. trachyrhynchus,*** Lath. White Pelican. Chiefly white, some black and yellowish; L. 60; W. 24; B. 12. N. Am., abundant S. and W., often inland.

FAMILY LXII.—PHALACROCORACIDÆ.

(*The Cormorants.*)

Bill about as long as head, nearly terete, strongly hooked, the cutting edges uneven; gular pouch small. Wings short. Tail very large, almost scansorial, of very stiff feathers, often used with the aid of the legs, which are set far back, as a support for the body; a nasal groove but nostrils abortive. Colors in both sexes lustrous, iridescent black; in the breeding season usually with long, white, filamentous plumes; many species crested. Genus one; species twenty-five; of most regions.

1. ***GRACULUS,*** Linnæus. CORMORANTS.

2. ***G. dilophus,*** (Sw.) Gray. DOUBLE-CRESTED CORMORANT. Glossy greenish black; back and wing coverts coppery gray; adult with two curly black lateral crests; sac convex or straight-edged behind, orange; L. 33; W. 13; T. 7; tail of 12 feathers. N. Am.; our commonest species.

2. ***G. mexicanus,*** (Brandt) Bon. MEXICAN CORMORANT. Gular sac orange, white-edged; L. 24. S. W., N. to Ills.

3. ***G. carbo,*** (L.) Gray. NORTHERN CORMORANT. Tail of 14 feathers; sac heart-shaped behind; L. 36. Northern and coastwise.

ORDER R.—LONGIPENNES.

(*The Long-Winged Swimmers.*)

Feet palmate; tibiæ feathered; legs near center of equilibrium; hind toe elevated, small, often wanting. Bill usually long, horny, not serrate nor lamellate; nostrils developed; no gular pouch. Wings very long and pointed. Tail well developed; altricial; powers of flight remarkable; food chiefly fishes. There are two families, *Laridæ*, below described, and *Procellaridæ*, the Petrels, which we here omit, all our numerous representatives being strictly maritime.

FAMILY LXIII. — LARIDÆ.

(*The Gulls.*)

Long-winged birds, with the nostrils not tubular; bill various. Hind toe small and elevated, but less so than in the Petrels. General color usually white, with a darker mantle of a pearly bluish tint, and commonly with some black markings. Sexes alike in color, but the plumage varying much with age and season. Genera about twelve (*Coues*), sixty (European authors); species ninety; abounding about all large bodies of water. Of the genera admitted below, *Pagophila*, *Rissa*, and *Chrœcocephalus* may properly be considered as sections of *Larus*, while *Gelochelidon* and *Thalasseus* are perhaps sub-genera under *Sterna*.

* Bill hooked (epignathous) — rarely not hooked, and the tail even.

† Bill with a sort of cere; middle tail feathers exserted; Jægers. (LESTRIDINÆ.) . . . . STERCORARIUS, 1.

†† Bill not cered; general color usually white with a darker mantle. Gulls. (LARINÆ.)

a. Hind toe rudimentary, without a developed claw. RISSA, 3.

aa. Hind toe perfect, provided with a claw.

b. Tail even.

c. Tarsus black, rough; webs incised; plumage white. PAGOPHILA, 4.

cc. Tarsus not black; lower plumage white in adult.

d. Head white—if dark below, head not whitish (species of large size; never rosy-tinted below; the head never with a dark hood.) . LARUS, 2.

dd. Head dark—if dark below, head whitish (small or slender species, with a black hood in the breeding season, and the white under parts then pinkish or rosy-tinted.) . . . CHRŒCOCEPHALUS, 5.

bb. Tail forked; bill black, yellow-tipped. . XEMA, 6.

** Bill paragnathous (mandibles even); tail forked (in our species); Terns. (STERNINÆ.)

e. Toes full-webbed; colors chiefly white, with a black cap at most seasons and the quills silvery dusky, with a long white stripe.

f. Feet black; forehead without white crescent.

g. Not crested; bill stout, scarcely longer than tarsus. GELOCHELIDON, 7.

gg. Crested; bill slender, much longer than tarsus. THALASSEUS, 8.

ff. Feet not black; back pale, no crest. . . STERNA, 9.

ee. Toes not full-webbed; color quite dark. HYDROCHELIDON, 10.

*** Bill hypognathous—the lower mandible much the longer, compressed like a knife-blade. Skimmers. (RHYNCHOPINÆ.) RHYNCHOPS, 11.

1. STERCORARIUS, Brisson. JÆGERS.

= *Lestris*, Authors.

1. ***S. pomatorhinus,*** (Temm.) Lawr. POMARINE JÆGER. Chiefly blackish, colors varying with age; middle tail feathers broad to the tip, projecting about four inches; L. 20; W. 15. Northern, U. S. in winter.

2. ***S. parasiticus,*** (Brünn.) Gray. PARASITIC JÆGER. General color dark brown; middle tail feathers acuminate, projecting 4 inches; L. 18; W. 13. Northern, U. S. in winter.

3. ***S. buffoni,*** (Boie.) Coues. LONG TAILED JÆGER. Similar, but still smaller; tail feathers filamentous, projecting 8 or 10 inches. Northern, U. S. in winter.

2. ***LARUS***, Linnæus. Gulls.

* Primaries without any black.

1\. ***L. glaucus***, Brünn. Glaucous Gull. Ice Gull. Burgomaster. Bill yellow with red spot on lower mandible; large; L. 30; W. 18. Arctic regions; S. in winter.

2\. ***L. leucopterus***, Faber. White-Winged Gull. Similar but smaller; L. 23; W. 17. Same region.

** Primaries crossed with black (adult), or all black (young).

3\. ***L. marinus***, L. Great Black-Backed Gull. Coffin-Carrier. Saddle-Back. Mantle blackish slate color; largest of our Gulls; L. 30 or more; W. 18; feet flesh colored. N. Atlantic, S. in winter.

4\. ***L. argentatus***, Brünn. Herring Gull. Common Gull. Mantle grayish blue; large, 22 to 27; W. 18 or less; feet flesh colored. N. Am., abundant.

5\. ***L. delawarensis***, Ord. Ring-Billed Gull. Plumage like preceding; feet olivaceous; webs yellow; bill yellowish, a black band at the tip; size moderate; L. 20; W. 15. N. Am., abundant.

3. ***RISSA***, Leach. Kittiwakes.

1\. ***R. tridactyla***, (L.) Bon. Kittiwake Gull. Mantle dark grayish blue; hind claw a minute knob; L. 16 to 18; W. 12. Northern, U. S. in winter.

3. ***PAGOPHILA***, Kaup. Ivory Gulls.

1\. ***P. eburnea***, (Gm.) Kaup. Ivory Gulls. Adults pure white; young spotted; L. 16 to 20; W. 12. Northern, rarely to U. S. in winter.

5. ***CHRŒCOCEPHALUS***, Eyton. Rosy Gulls.

1\. ***C. atricilla***, (L.) Lawr. Black-Headed or Laughing Gull. Tarsus $\frac{1}{4}$ longer than middle toe and claw;

large; bill and feet dusky carmine; L. 16 to 19; W. 12 to 13. U. S., coastwise.

2. ***C. franklini,*** (Rich.) Bruch. Franklin's Rosy Gull. Tarsus about as long as middle toe and claw; bill and feet carmine; bill usually with a black mark; medium; L. 14 to 16; W. 11. U. S., chiefly W. of the Miss. R.

3. ***C. philadelphia,*** (Ord.) Lawr. Bonaparte's Gull. Tarsus about as long as middle toe and claw; bill dark or black, slender, tern-like; small; L. 12 to 14; W. 10. N. Am., abundant.

6. *XEMA,* Leach. Fork-Tailed Gulls.

1. ***X. sabinei,*** (Sab.) Leach. Forked-Tail Gull. Chiefly white, a black hood and collar; L. 14; W. 11. Northern, S. in winter to N. Y.

7. *GELOCHELIDON,* Brehm. Gull-Billed Terns.

1. ***G. anglica,*** (Montagu.) Bon. Marsh Tern. Bill black, very short and stout; L. 15; W. 12. E. U. S., not abundant.

8. *THALASSEUS,* Boie. Crested Terns.

1. ***T. caspius,*** (Pallas.) Boie. Caspian Tern. Primaries without white band; bill red; much the largest of the Terns; L. 20 or more; W. 17; T. 6, not much forked. Northern, S. in winter; scarce.

2. ***T. regius,*** Gamb. Royal Tern. Bill orange; L. 18 or 20; W. 15; T. 8, deeply forked; much smaller than the last but nearly as long. Atlantic Coast.

3. ***T. cantiacus,*** (Gm.) Boie. Sandwich Tern. Bill black, yellow at tip; L. 16; W. 12½; T. 6. Europe and Am.; rare on our coast.

9. *STERNA*, Linnæus. COMMON TERNS.

1. ***S. hirundo***, Auct. COMMON TERN. SEA SWALLOW. WILSON'S TERN. Bill red, blackening towards tip; tail mostly white; outer web of outer feather darker than inner; L. 14½ (13 to 16); W. 10 (9½ to 11¾); T. 6 (5 to 7.) Coasts of Europe and America; abundant. (*S. wilsoni*, Lawr.)

2. ***S. forsteri***, Nuttall. FORSTER'S TERN. Larger; tail longer and wings shorter; inner web of outer tail feather darker; W. 9½ to 10½; T. 6½ to 8. N. Am., common.

3. ***S. macrura***, Naumann. ARCTIC TERN. Bill carmine throughout; plumage as in *hirundo*, but darker below; L. 14 to 17; W. 10 to 12; T. 5 to 8; smaller than *hirundo*, but tail proportionally much longer. Northern regions, S. to U. S.

4. ***S. paradisea***, Auct. ROSEATE TERN. Bill black, usually orange at base below; mantle very pale; somewhat rosy-tinted below; L. 12 to 16; W. 9 to 10; T. 5 to 8. Atlantic Coast, abundant. (*S. dougalli*, Mont.)

5. ***S. portlandica***, Ridgway. PORTLAND TERN. Near the preceding, but mantle as in *hirundo;* the rump white instead of pearly; feet blackish; under parts pure white; L. 12½; W. 9¾; T. 5 or more. Lately discovered in Maine and Mass.; but two specimens known.

6. ***S. superciliaris***, (Vieill.) var. ***antillarum***, (Lesson) Coues. LEAST TERN. Bill yellow, usually tipped with black; a white frontal crescent between cap and bill; shafts of two or more outer primaries black above; very small; L. 8 or 9; W. 6¼; T. 2 to 3½. U. S., chiefly abundant coastwise.

10. HYDROCHELIDON, Boie. BLACK TERNS.

1. ***H. lariformis,*** (L.) Coues. BLACK TERN. Head, neck and under parts black (in full plumage); wings and tail above plumbeous like the back; crissum white; small; L. 10; W. 8 to 9; T. 3½. N. Am., chiefly inland. [*H. fissipes*, (L.) Gray.]

2. ***H. nigra,*** (L.) Gray. WHITE-WINGED BLACK TERN. Wings whitening along border of fore-arm; tail and upper tail coverts white. Straggler from Europe, a single specimen lately taken on Lake Koshkonong. (*Ludovic Kumlien.*) [*H. leucoptera*, (Meisn.) Boie.]

11. RHYNCHOPS, Linnæus. SKIMMERS.

1. ***R. nigra,*** L. BLACK SWIMMER. CUTWATER. Glossy black; white below; lower mandible about an inch longest, compressed like a knife-blade, obtuse at end; L. 16 to 20; W. 15; T. 5, sharply forked. Coast, abundant southward.

ORDER S.—PYGOPODES.

(*The Diving Birds.*)

Feet palmate or lobate; tibiæ feathered, buried in the skin nearly to the heel joint, hence the legs are set very far back, and the birds are scarcely able to walk at all on land; hind toe small, elevated, often wanting. Nostrils developed; bill of various forms, horny, not lamellate nor serrate; no gular pouch. Wings very short, scarcely reaching the base of the very small or rudimentary tail. Swimmers, many of them noted for their powers of diving. We here omit the three-toed family of ALCIDÆ, the Auks, they being strictly maritime and mostly northern. The twenty-one known species all occur in America.

FAMILY LXIV.—COLYMBIDÆ.

(*The Loons.*)

Bill long, strong, tapering, acute, wholly hard; nostrils linear. Head densely and evenly feathered, without ruffs or naked spaces; eye large. Feet 4-toed, palmate; tarsus reticulate, strongly compressed. Wings comparatively long and strong. Tail short, but well developed. Back of adult with small spots; precocial. Genus one; species three. Birds of large size, with strong powers of flight, and pre-eminent in swimming and diving, but scarcely able to walk; they are migratory, breeding northward, but coming S. in winter; the voice is singularly sharp and wild.

1. **COLYMBUS,** Linnæus. LOONS.

= *Eudytes*, Illiger.

1. ***C. torquatus,*** Brünn. GREAT NORTHERN LOON. DIVER. Black; breast and below chiefly white; head and neck iridescent, green and violet; a patch of white streaks on each side of neck and on the throat; back with many white spots; L. 36; W. 14; Ts. 3; B. 3, Northern Hemisphere; whole U. S. in winter. (*C. glacialis*, L.)

2. ***C. arcticus,*** L. BLACK-THROATED DIVER. Similar, but head and neck behind bluish or hoary gray; fore-neck purplish black, with a crescent of white streaks; L. 28; W. 12; B. 2½. Northern hemisphere, not common in U. S.

3. ***C. septentrionalis,*** L. RED-THROATED DIVER. Blackish, chiefly white below; head and neck mostly bluish gray; throat with a large chestnut patch; L. 27; W. 11; B. 2. Northern hemisphere.

FAMILY LXV.—PODICIPIDÆ.

(*The Grebes.*)

Bill usually slender, rarely stoutish; lores naked; head often with crests, ruffs or ear tufts in the breeding season. Back not spotted; under plumage lustrous, mostly white. Wings very short. Tail rudimentary. Feet four-toed, lobate, the toes webbed at base; toes flattened, provided with flat claws resembling human nails; tarsus scutellate, compressed. Genera two; species about twenty; in all parts of the world, chiefly about fresh waters.

* Bill slender, straight, rather acute; loral strip narrow; head in breeding season with conspicuous crests or ruff. PODICEPS, 1.

** Bill stout, somewhat hooked; loral strip broad; no ruff nor crest. . . . . PODILYMBUS, 2.

1. PODICEPS, Latham. CRESTED GREBES.

= *Colymbus*, Illiger.

1. ***P. cristatus,*** (L.) Lath. CRESTED GREBE. Upper parts generally dark brown; crest black; throat and sides of head white, becoming reddish on the ruff; primaries brown; secondaries mostly white; silky white below, not mottled; L. 24; W. 8½; B. 2. Northern hemisphere; U. S. in winter.

2. ***P. holbolli,*** Reinhardt. RED-NECKED GREBE. Upper parts brown; front and sides of neck rich brownish red; throat and sides of head ashy; crests and ruffs not large; below silvery ash, spotted or mottled; L. 18; W. 8; B. 1¾. N. Am., U. S. in winter.

3. ***P. cornutus,*** (Gm.) Lath. HORNED GREBE. Dark brown; head glossy black; a brownish yellow band over eye and behind; fore-neck and breast brownish red; bill

compressed, black, tipped with yellow; crests and ruffs very large; L. 14; W. 6; B. $3\frac{3}{4}$. Northern hemisphere, abundant.

4. ***P. auritus,*** (L.) var. ***californicus,*** (Heerm.) Coues. EARED GREBE. Crest in the form of ear tufts; front of neck black; bill depressed; L. 12. Western, E. to Ills.

2. PODILYMBUS, Lesson. DAB-CHICK.

1. ***P. podiceps,*** (L.) Lawr. DIEDAPPER. HELL-DIVER. WATER WITCH. PIED-BILLED GREBE. Chiefly brownish gray; silvery ash below; bill bluish, with dark band; young and winter plumage different, but the bird resembles nothing else; L. 14; W. 5; B. 1. Whole of America, abundant. (*Podiceps carolinensis*, Lath.)

Class III. — Reptilia.

(*The Reptiles.*)

A Reptile is a cold-blooded, air-breathing vertebrate, having the exoskeleton developed as horny or bony plates, never as feathers or hair. Limbs, when present, usually adapted for walking, rarely for swimming, scarcely ever for flying. An incomplete double circulation, the ventricular septum being usually imperfect or wanting; no metamorphosis; oviparous, rarely ovoviviparous, the eggs relatively large and usually with a leathery skin. Various important anatomical distinctions exist, but the Reptiles are obviously separated from the Birds by the absence of feathers, and from the Batrachians by the presence of scales, and by the absence of gills after leaving the egg.

Besides the three following orders, a fourth (CROCODILIA), is represented by two species *Alligator mississippiensis*, Daudin, and *Crocodilus americanus*, Seba, in our Southern States.

ORDERS OF REPTILIA.

* Body covered with square imbedded shields; vent roundish or longitudinal, plaited; bones of skull soldered together.

† Body short, depressed, enclosed between two bony shields, from which the head, limbs and tail may be protruded; no teeth. . . . . TESTUDINATA, T.

** Body covered with imbricated scales; vent a cross-slit; bones of skull separate; jaws with teeth.

‡ Mouth not dilatable; bones of mandible united by a bony suture in front; limbs 4—rarely rudimentary. LACERTILIA, U.

‡‡ Mouth very dilatable; bones of mandible united by ligaments; limbs wanting or represented only by short spurs on the sides of the vent. . . . . OPHIDIA, V.

T. FAMILIES OF TESTUDINATA.

* Carapace firm, not flexible at the margins, not greatly depressed; both shields with well-developed horny plates.

† Toes short, bound together by the integument; legs and feet short, club-shaped; carapace very convex; plastron covering nearly all of under surface of body; caudal shields united; claws blunt, 5–4; terrestrial. TESTUDINIDÆ, 66.

†† Toes well developed, spreading, and in aquatic species webbed; claws usually 5–4.

‡ Shell highest at about the middle, usually somewhat depressed, the margin flaring outwards; epidermal plates of the large plastron 12 in number. . EMYDIDÆ, 67.

‡‡ Shell highest behind the middle; margin of carapace turned rather downward or inward; plates of plastron 7, 9 or 11—never 12; size small. . CINOSTERNIDÆ, 68.

‡‡‡ Shell highest anteriorly; carapace flaring outward, its margin toothed behind; plastron small, cross-shaped with 12 plates and three accessory ones on each side; jaws powerful, strongly hooked; neck and tail long, the latter with a crest of tubercles; size large. CHELYDRIDÆ, 69.

** Much depressed; carapace and plastron covered with a leathery skin, and flexible at the margins; no horny plates; fleshy lips; snout prolonged; toes 5–5, but claws 3–3. TRIONYCHIDÆ, 70.

U. FAMILIES OF LACERTILIA.

* Tongue thick, convex, attached at its base to the gullet; scales usually more or less spinous. . . . IGUANIDÆ, 71.

** Tongue flat, elongate, bifid at the end; scales never spinous.

† Limbs rudimentary, concealed beneath the skin; sides with a longitudinal fold. . . . . ANGUIDÆ, 72.

†† Limbs four—well developed.

‡ Scales of the belly rounded, arranged in quincunx order. SCINCIDÆ, 74.

‡‡ Scales of the belly quadrate, arranged in cross-bands; throat with two cross-folds. . . . TEIDÆ, 73.

V. FAMILIES OF OPHIDIA.

* Both jaws fully provided with small teeth; no poison fangs; no rattle; no anal appendages; no ante-orbital pit; *not venomous*. . . . . COLUBRIDÆ, 75.

** Upper jaw with enlarged, erectile poison fangs, otherwise toothless; a deep pit between eye and nostril; *venomous*. CROTALIDÆ, 76.

*** Upper jaw with small, permanently erect poison fangs; no ante-orbital pit; color red, with black rings; somewhat *venomous*. . . . . ELAPIDÆ, 75. (b.)

ORDER T.—TESTUDINATA.

(*The Turtles.*)

Reptiles with the body enclosed between two more or less developed bony shields, which are usually covered by horny epidermal plates, but sometimes (*Trionychidæ*, *Sphargididæ*) by a leathery skin. The carapace (upper shield) and plastron (lower shield) are more or less united along the sides. The neck and the tail are the only flexible parts of the spinal column, and these, together with the legs, can usually be retracted within the box made by the two shields. The bony part of the carapace is formed by the dorsal and sacral vertebræ, and the ribs co-ossified with a series of overlying bony plates, usually accompanied by a marginal row. The dorsal vertebræ have their ends flattened and immovably united by cartilage, and all of them, except the first and last, have their neural spines flattened horizontally so as to form the median line of plates. On either side of this series is a single row of ossified dermal plates overlying the ribs and corresponding in number to the developed ribs of which there are usually eight pairs.

No traces of a true sternum have been discovered (*Huxley*). The plastron consists of membrane bones, of which there are usually nine pieces — four pairs and a single symmetrical median piece. These correspond neither in number nor position with the overlying dermal plates.

The skull is more compact than that of the other Reptiles. There are no teeth, but the jaws are encased in horny sheaths, usually with sharp cutting edges; the eye is furnished with two lids and a nictitating membrane as in the Birds; the tympanic membrane is always present, although sometimes hidden by the skin. Respiration is effected by swallowing air.

The order *Testudinata* is divided by Prof. Agassiz into two sub-orders: — AMYDÆ, comprising the Land and Fresh Water Turtles, with retractile feet that may be used for walking; and CHELONII, the Sea Turtles, with flipper-like feet used chiefly for swimming. Of the latter, several species occur on our coast, but we here omit them.

FAMILY LXVI. — TESTUDINIDÆ.

(*The Land Tortoises.*)

Carapace strong, thick, ovate, generally very convex and falling off abruptly at both ends; caudal shields united into one; plastron very broad, covering the whole under surface, the anterior part sometimes movable on a transverse hinge. Legs and feet club-shaped; toes firmly bound together by the integument, only the blunt claws being exserted.

Herbivorous Turtles, entirely terrestrial, inhabiting the warmer parts of both continents; about twenty species are known.

1. TESTUDO, Linnæus. LAND TORTOISES.

> *Xerobates*, Agassiz.

1. ***T. carolina,*** L. CAROLINA "GOPHER." L. 15. S. States, N. to N. C.; burrows in the ground like a woodchuck.

FAMILY LXVII.—EMYDIDÆ.

(*The Pond Turtles.*)

Carapace ovate, broadest behind, the margin having a tendency to flare outward, highest near the middle, usually rather depressed, rarely strongly convex; plastron covering the whole under surface, its plates twelve in number; sometimes the anterior lobe (and rarely the posterior also) movable on a transverse hinge, enabling the animal to completely close the shell. Toes broadly webbed in the aquatic species; scarcely webbed in the others. Jaws never hooked and pointed, as in allied families. They feed largely upon animals, but they rarely catch active prey. They do not bite except under much provocation. Species seventy or eighty, widely distributed, inhabiting marshes, ponds, and the shores of still streams; a few are strictly terrestrial.

* Carapace short, very high and strongly convex; plastron united to the carapace by a more or less cartilaginous suture and divided by a transverse hinge into two or more movable pieces; the anterior one, the smaller; toes scarcely webbed; terrestrial. . . . . CISTUDO, 1.

** Carapace somewhat elongated, considerably arched; plastron immovable; toes short, with a small web; feet more nearly equal, and habits less aquatic than in the succeeding groups; species of small size.

a. Shell more or less carinated, without round spots; upper jaw deeply notched and arched downward. . CHELOPUS, 2.

aa. Shell not carinated, black, usually with round, yellowish spots; upper jaw slightly notched, its edges nearly straight. NANEMYS, 3.

*** Carapace rather depressed; plastron wide, flat, movable upon the carapace and also upon a transverse hinge; anterior lobe somewhat smaller than the posterior, which is emarginate behind; toes webbed. . . . . EMYS, 4.

**** Carapace rather flat; plastron wide and flat, as is also the bridge connecting it to the carapace; toes broadly webbed; hind-legs much stouter than fore-legs; larger species, decidedly aquatic.

† Upper jaw not notched in front; carapace more or less strongly keeled or tuberculated. . . . MALACOCLEMMYS, 6.

†† Upper jaw notched in front; shell not keeled in adult.

‡ Horizontal alveolar surfaces of jaws not divided by a longitudinal ridge; stripes on neck, tail, legs, etc., bright red (in our species); head with yellow lines; large plates of carapace plain; marginal plates with bright red markings; a small tooth on each side of notch in upper jaw; shell never keeled. . . . . CHRYSEMYS, 5.

‡‡ Horizontal alveolar surfaces divided by a longitudinal ridge, running parallel with the cutting edge; stripes on legs, etc., usually yellow, never bright red; large plates of carapace often variegated; traces of a keel usually evident, at least in the young. . . PSEUDEMYS, 7.

1. CISTUDO, Fleming. Box Turtles.

1. **C. clausa,** (Gm.) COMMON BOX TURTLE. Colors very variable, chiefly blackish variegated with yellowish; N. Y. to Mo. and S. in dry woods.

Var. **triunguis,** (Ag.) Cope. THREE-TOED BOX TURTLE. Hind-feet mostly 3-toed, paler. Southern, N. to Penn.

2. **C. ornata,** Ag. NORTHERN BOX TURTLE. "Shell round, broad, flat, without keel, even when young." Iowa and W.

2. CHELOPUS, Rafinesque. Wood Turtles.

* A deep notch in upper jaw, with a lengthened tooth on each side of it; lower jaw strongly arched upwards. (*Calemys*, Ag.)

1. ***C. muhlenbergii,*** (Schweigger) Cope. MUHLENBERG'S TORTOISE. Brown with yellowish markings; plastron black with yellowish central blotch; an orange spot on each side of neck; shell somewhat carinated; L. 4½. E. Penn. and N. J.

** Upper jaw broad at end, arched downward, with a notch at tip; just behind the tip the horny sheath slants inward so that the width of the jaw is less than that of the forehead; edge of lower jaw straight, excepting the tip which is strongly upcurved. (*Glyptemys*, Ag.)

2. ***C. insculptus,*** Le C. WOOD TORTOISE. Shell carinated, its plates marked with concentric striæ and radiating lines; plastron with a black blotch on each plate; L. 8. U. S., E. of Ohio, in woods and fields.

3. *NANEMYS,* Agassiz. SPECKLED TORTOISES.

1. ***N. guttatus,*** (Schn.) Ag. SPECKLED TORTOISE. Black, dotted more or less with orange, these spots rarely obsolete; plastron yellow, blotched with black; shell not carinated; L. 4½. E. U. S., W. to N. Ind. (*Levette*); abundant.

4. *EMYS,* Brogniart. TORTOISES.

1. ***E. meleagris,*** (Shaw) Ag. BLANDING'S TORTOISE. Jet black; usually with yellowish spots; plastron yellowish with black blotches; head with yellow spots; L. 8. Wisconsin to Alleghanies, in moist woods and fields.

5. *CHRYSEMYS,* Gray. PAINTED TURTLES.

1. ***C. picta,*** (Herm.) Ag. PAINTED TURTLE. MUD TURTLE. Greenish black; plates margined with paler; marginal plates marked with bright red; plastron yellow, often blotched with brown; L. 8. E. U. S., one of the most common turtles.

Var. ***marginata,*** Ag. Plates of carapace alternating or in quincunx, the lateral rows out of line with the middle one, instead of forming sets of three as in the eastern form; lateral plates with strong concentric striæ. W. N. Y. and W., common. *C. oregonensis*, (Holb.) Ag., without red markings, occurs in Minn. and W.

6. MALACOCLEMMYS, Gray. MARSH TURTLES.

* Lower jaw spread out into a spoon-shaped dilatation; head with a horny skin; inland turtles. (*Graptemys*, Ag.)

1. ***M. geographicus,*** (LeS.) Cope. MAP TURTLE. Dark olive brown with greenish and yellow streaks and reticulations, especially distinct on neck, legs and edges of carapace; plastron yellowish; carapace strongly notched behind and usually decidedly keeled. Miss. Valley, E. to N. Y.

2. ***M. pseudogeographicus,*** (Holbr.) Cope. LESUEUR'S MAP TURTLE. Much like the preceding but grayer, the markings on the shell paler, less distinct and in larger pattern; keel of carapace stronger, each plate of the vertebral series with a blackish projection behind, which is more or less imbricated over the succeeding plate; plastron yellowish, marbled with blackish; head, neck and legs with bright yellow stripes. Wis. to Ohio and S. W.

** Sheath of jaws straight, the cutting edges smooth; salt-marsh turtles. (*Malacoclemmys.*)

3. ***M. palustris,*** (Gmel.) SALT-MARSH TURTLE. DIAMOND-BACK. Greenish or dark olive; plates, both of carapace and plastron, with concentric dark stripes. N. Y. to Texas, along the coast.

7. PSEUDEMYS, Gray. TERRAPINS.

* Jaws serrated. (*Ptychemys*, Ag.)

1. ***P. rugosa,*** (Shaw). RED-BELLIED TERRAPIN. Dusky, with red markings above; marginal plates with much red; plastron red or partly yellowish; neck, etc., with

yellow stripes; variable; N. J. to Va., an elegant turtle, known by the serrated jaws.

** Jaws not serrated. (*Trachemys*, Ag.)

2. ***P. hieroglyphica,*** (Holbr.) HIEROGLYPHIC TURTLE. Shell smooth, depressed; olive brown with broad reticulated, yellowish lines; plastron dingy yellow; head very small. E. U. S.

3. ***P. troostii,*** (Holbr.) YELLOW-BELLIED TERRAPIN. Greenish-black, lateral plates with horn-colored lines and spots; plastron dull yellow, with large, black blotches; throat with greenish stripes; shell never keeled. Miss. Valley, N. to Ills.

4. ***P. elegans,*** (Wied.) ELEGANT TERRAPIN. Brown with yellowish wavy lines and blotches; a blood-red band on each side of neck; plastron yellow with a dusty blotch on each plate. Ills. to Rocky Mountains.

5. ***P. scabra,*** (L.) Cope. ROUGH TERRAPIN. Dark brown, with yellow stripes; plastron yellow with small black blotches in front; carapace wrinkled. Va. to Fla.

FAMILY LXVIII.—CINOSTERNIDÆ.

(*The Cinosternoid Turtles.*)

Carapace rather long and narrow, the outline usually rising gradually from the front to a point beyond the center of the shell, then abruptly descending; the bulk of the body therefore thrown backward; margin of the carapace turning downward and inward rather than outward; plastron proportionally large, covered with 7, 9 or 11 horny plates, the anterior pair coalescing into one; anterior, and sometimes also posterior lobe of plastron, often movable upon the fixed central portion; head pointed; jaws usually strong.

Turtles of small size, chiefly American.

* Anterior and posterior lobes of plastron nearly equal, both freely movable and capable of closing the shell; posterior lobe emarginate behind, its angles rounded; carapace without traces of keel in adult. . . . . CINOSTERNUM, 1.

** Posterior lobe of plastron narrower and longer, truncate behind, its angles rather pointed; lobes of plastron little movable, incapable of closing the shell; carapace more or less carinated, at least when young; head very large, with strong jaws. AROMOCHELYS, 2.

1. CINOSTERNUM, Wagler. SMALL BOX TURTLES.

> *Thrynosternum*, Ag.

1. **C. pennsylvanicum,** (Bosc.) Bell. SMALL MUD TURTLE. Shell dusky brown; head and neck with light stripes and yellow dots; anterior dorsal plate nearly as broad in front as long; L. 4. N. Y. to Fla. and W.

2. AROMOCHELYS, Gray. MUSK TURTLES.

= *Ozotheca*, Agassiz.

1. **A. odoratus,** (Latreille) Gray. MUSK TURTLE. STINK-POT. Shell dusky, clouded, sometimes spotted; head very large with strong jaws; carapace with traces of a keel, but the plates not imbricated in the adult; anterior dorsal plate nearly twice as wide as long in front; a yellow stripe from snout, above eye, down the side of neck and another below eye; a strong musky odor; L. 6. E. U. S., abundant. W. to Indiana. (*Levette.*)

2. **A. carinatus,** Gray. LITTLE MUSK TURTLE. Plates of carapace overlapping more or less, each one edged with black and marked with radiating stripes; neck unstriped. Lower Mississippi region. (*Goniochelys minor*, Ag.)

FAMILY LXIX. — CHELYDRIDÆ.

(*The Snapping Turtles.*)

Shell high in front, low behind; bulk of body thrown

forward; head and neck very large; jaws strongly hooked, and exceedingly powerful; tail long, strong, with a crest of horny, compressed tubercles; plastron small, cross-shaped, covered with twelve plates; bridge very narrow.

Large turtles of great strength and voracity, chiefly aquatic; two of the three species are American, the third (*Platysternum*) is from China. Their fierceness is well known; when angry they elevate the body, and, in biting, throw themselves forcibly forward.

* Head rough, covered with soft skin; tail with two rows of large scales beneath; ridges of carapace disappearing with age; jaws moderately hooked. . . . CHELYDRA, 1.

** Head very large, covered with smooth, symmetrical plates; tail with many small imbricate scales beneath; carapace very strongly three-keeled; jaws very strongly hooked. MACROCHELYS, 2.

1. CHELYDRA, Schweigger. SNAPPING TURTLES.

1. **C. serpentina,** (L.) Schw. COMMON SNAPPING TURTLE. Canada to Equador, every where abundant.

2. MACROCHELYS, Gray. ALLIGATOR SNAPPERS.

= *Gypochelys*, Ag.

1. **M. lacertina,** (Schw.) MISSISSIPPI SNAPPER. Gulf States, N. to Illinois; "perhaps the most ferocious, and, for their size, the strongest of reptiles."

FAMILY LXX.—TRIONYCHIDÆ.

(*The Soft-Shelled Turtles.*)

Body flat, nearly orbicular; carapace not completely ossified, the ribs projecting freely towards the outer extremities; marginal ossicles rudimentary; carapace and plastron covered by a thick leathery skin which is flexible at the margins. Head long and pointed with a long, flexible, tubular, pig-like snout; neck long. Feet broadly webbed; toes long, 5–5, but the claws only 3–3.

Aquatic, carnivorous and voracious; species about 30, in both hemispheres.

* Nostrils terminal, crescent-shaped; a prominent longitudinal ridge projecting from each side of septum. ASPIDONECTES, 1.

** Nostrils rather under the tip of snout; nasal septum without an internal longitudinal ridge on each side. AMYDA, 2.

1. ASPIDONECTES, Wagler. SOFT-SHELLED TURTLES.

1. ***A. spinifer,*** (LeSueur) Ag. COMMON SOFT-SHELLED TURTLE. Carapace olive brown with dark spots; plastron nearly white; head and neck olive green with light and dark stripes; legs and feet mottled every where with dark; male with the tubercles on the front of the carapace smaller than in the female, the body also longer and the tail extending considerably beyond the margin of the carapace. Great Lakes and Upper Mississippi, abundant.

2. ***A. nuchalis,*** Ag. CUMBERLAND TURTLE. A marked depression on either side of the keel, which is dilated and triangular anteriorly; spines and tubercles on carapace largely developed. Cumberland and Upper Tenn. Rivers.

2. AMYDA, Agassiz. LEATHERY TURTLES.

1. ***A. mutica,*** (Les.) Ag. LEATHERY TURTLE. A depression along median line of carapace; no spines nor tubercles along anterior margin nor on back; feet not mottled below. Upper Mississippi and Great Lakes.

ORDER U.—LACERTILIA.

(*The Lizards.*)

Reptiles not shielded, with the body usually covered

with overlapping scales; mouth not dilatable; tongue free; jaws always with teeth. Limbs four, distinct, rarely rudimentary and hidden by the skin; a pectoral arch developed. Feet usually with five digits, the phalanges normally 2, 3, 4, 5, 3 or 4. Tail usually long and in many cases very brittle, readily broken by a slight blow; this is owing to a thin, unossified, transverse septum, which traverses each vertebra. "The vertebra naturally breaks with great readiness through the plane of the septum, and when such lizards are seized by the tail, that appendage is pretty certain to part at one of these weak points" (*Huxley*). Vent a cross slit; urinary bladder present. The great majority of the numerous species belong to tropical and sub-tropical regions.

FAMILY LXXI.—IGUANIDÆ.

(*The Iguanas.*)

New World lizards of various habits; the tongue short and thick and the eyes diurnal with round pupils; scales imbricated, those on the belly small and rhombic. Feet for walking; toes unequal. Tail with more or less distinct whorls of scales, which are commonly spinous. Warmer parts of America. Genera about sixty; species one hundred and fifty, or more. (*Gray.*)

* Body moderately depressed; head broad, not spinous; ventral plates not keeled. . . . . . . SCELOPORUS, 1.

** Body much depressed; head armed with stout spines. PHRYNOSOMA, 2.

1. ***SCELOPORUS,*** Wiegmann. TREE SWIFTS.

= *Tropidolepis*, Cuvier.

1. ***S. undulatus,*** (Harlan.) PINE TREE LIZARD. SWIFT. Greenish, bluish, or bronzed, with black, wavy cross bands above; throat and sides of belly usually with

8

brilliant blue or green; dorsal scales rather large, carinated; tail slender; L. 7. U. S., in pine forests, etc.; abundant southward; varies greatly in color.

2. PHRYNOSOMA, Wiegmann. Horned Toads.

1. ***P. douglasi,*** Bell. Horned Toad. Teguexin. No row of large spines along sides of belly; ventral scales smooth. Central and Western parts of U. S. southwestward occurs *P. cornutum* (Harlan) Gray, with a row of stout lateral spines, and carinated ventral plates.

FAMILY LXXII.—ANGUIDÆ.

(*The Glass Snakes.*)

Limbs rudimentary, hidden under the skin; body, therefore, snake-like in form, but the general aspect lizard-like. Through ignorance of the boundaries of this family, I am unable to characterize it.

1. OPHEOSAURUS, Daudin. Glass Snakes.

1. ***O. ventralis,*** (L.) Daud. Glass Snake. Body serpentiform; a conspicuous lateral fold; no external limbs; tail very brittle, as in most lizards; dusky and yellow with narrow black streaks. Tenn. to Kansas and S.

FAMILY LXXIII.—TEIDÆ.

(*The Teguexins.*)

New World Lizards; head pyramidal, with regular many-sided shields; scales of back granular or carinated; throat scaly, usually with a double collar. Warmer parts of America; genera 12; species about 40.

1. CNEMIDOPHORUS, Wiegmann. Taraguiras.

1. ***C. sexlineatus,*** (L.) D. & B. Six-Lined Lizard. Olive, with 3 or 4 yellow streaks on each side; abdomen

silvery; length 6 to 9 inches. S. E. Va. to Ills. and Mexico.

FAMILY LXXIV.—SCINCIDÆ.

(*The Skinks.*)

Head subquadrangular; regularly shielded; body fusiform or subcylindrical, often with longitudinal stripes; limbs 4, various. Genera 50; species 150; in most parts of the world.

* Ear large; its front edge dentate; lower eye-lid scaly. EUMECES, 1.

** Ear very large, circular, its front edge rounded; lower eye-lid with a transparent disk. . . . . OLIGOSOMA, 2.

1. *EUMECES*, Wiegmann. BLUE TAILS.

= *Plestiodon*, Auct.

1. ***E. fasciatus***, (L.) BLUE-TAILED LIZARD. Blackish, with fine yellowish streaks, middle one forked on the head; tail mostly blue; old specimens sometimes reddish olive, obscurely striped; head reddish; L. 8 to 11. U. S., E. of the Rocky Mts.; abundant S.; very variable.

2. ***E. septentrionalis***, (Baird) Cope. NORTHERN SKINK. Olive, with four dark stripes above; sides with two narrow white lines margined on each side with black. Minnesota to Nebraska.

3. ***E. anthracinus***, (Baird) Cope. COAL SKINK. Black (?). Alleganies, from Penn. S.

2. *OLIGOSOMA*, Girard. MOCOS.

1. ***O. laterale***, (Say.) Grd. GROUND LIZARD. Chestnut color; on each side a black lateral band, edged with white; abdomen yellowish; tail blue below; head short; small and slender; L. 5. Southern States, abundant; N. to Illinois. (*Nelson.*)

ORDER V.—OPHIDIA.

(*The Serpents.*)

Reptiles, not shielded, with an epidermal covering of imbricated scales, which is shed as a whole and replaced at regular intervals. Mouth very dilatable, the bones of the lower jaw separate from each other, only united by ligaments. Limbs wanting, or represented by small spurs on the sides of the vent; vent a transverse slit. Various anatomical characters distinguish the snakes, but the elongated form and absence of limbs separate them at once from all our other Vertebrates, excepting the Lizard *Opheosaurus*, and this is not in any other respect, snake-like.

FAMILY LXXV. (*a*)—COLUBRIDÆ.

(*The Colubrine Snakes.*)

Both jaws fully provided with teeth, which are conical and not grooved; head covered with shields; no poison fangs; no spur-like appendages to vent; belly covered with broad band-like plates (gastrosteges); tail conical, tapering; sub-caudal plates (urosteges) arranged in pairs.

A very large family comprising nearly one hundred genera, and upwards of four hundred species, found in nearly every part of the world, but most abundant in warm regions. They differ from the *Elapidæ* of the Southern U. S. and southward, in the want of erect poison fangs; from the *Crotalidæ*, in having both jaws fully provided with teeth, and the absence of erectile poison fangs; and from the *Boidæ* and their relatives in the want of the spur-like rudimentary posterior limbs.

The following key is entirely artificial, and in the consideration of the species, I have generally omitted characters taken from the *cephalic* plates as not available for my purpose.

* Dorsal scales carinated.

† Anal plate entire.

a. Rows of scales 19 to 21; ventral plates (gastrosteges) 140 to 170; general color 3 light stripes on a darker ground; sides usually with spots; mostly viviparous.
EUTÆNIA, 5.

aa. Rows of scales 29 to 37; gastrosteges 200 to 240; general color whitish with a triple series of dark blotches.
PITYOPHIS, 7.

†† Anal plate bifid.

b. Rows of scales 15 to 17.

c. Tail about one-third of total length; gastrosteges 150 to 160; urosteges 100 or more; color clear golden green.
CYCLOPHIS, 9.

cc. Tail much less than one-third of total length; gastrosteges 125 to 130; urosteges 60 or less; color brown or reddish.

d. With one or two faint grayish dorsal stripes and usually a broken dusky band along sides; small species: rows of scales 15 to 17; loral plate absent; anteorbital present. . . . STORERIA, 4.

dd. No dorsal stripe; rows of scales 17; loral plate present; anteorbital wanting. . HALDEA, 16.

bb. Rows of scales 19 to 21.

e. Gastrosteges 130 to 170; general color usually 3 or more dark bands on a lighter ground; size large.
TROPIDONOTUS, 2.

ee. Gastrosteges 130 to 140; general color brown with many obscure black cross-blotches, distinct on neck; belly salmon red with a row of black spots on each side; size small. . . . . TROPIDOCLONIUM, 3.

bbb. Rows of scales 23 to 29.

f. Form short and stout; snout prominent, acute, recurved and keeled, forming a sharp ridge; head flattened when angry; gastrosteges 125 to 150. HETERODON, 1.

ff. Snout not recurved and keeled; species of large size.

g. Gastrosteges 130 to 155; general color 3 series of dark blotches on a lighter ground, sometimes simply barred or uniformly dark or reddish; aquatic. TROPIDONOTUS, 2.

gg. Gastrosteges 200 to 235; general color brown or black, sometimes with quadrate blotches; abdomen usually blotched; terrestrial. . . . COLUBER, 8.

** Dorsal scales smooth.

‡ Anal plate entire; gastrosteges 175 to 200· color black, brown or red, more or less variegated.

h. Rows of scales 21 to 25; loral plate present. OPHIBOLUS, 12.

hh. Rows of scales 19; no loral plate. . . OSCEOLA, 13.

‡‡ Anal plate bifid; scales in 13 to 17 rows.

i. Gastrosteges 170 to 210; scales in 17 rows; snakes very large or very long — ours lustrous pitch black in color when adult. . . . . . . BASCANIUM. 6.

ii. Gastrosteges less than 170; snakes of medium to small size.

j. Scales in 13 rows; gastrosteges 120 to 135; brown, salmon color or yellowish beneath. . CARPHOPHIOPS, 18.

jj. Scales in 15 to 17 rows.

k. Color black, unstriped; a distinct yellowish ring about neck; gastrosteges 140 to 160. . DIADOPHIS, 11.

kk. Color clear bright green; no collar; gastrosteges 130 to 140. . . . . LIOPELTIS, 10.

kkk. Color brownish, with black dots; no collar; gastrosteges 115 to 125. . . . . VIRGINIA, 17.

‡‡‡ Anal plate bifid; scales in 19 rows; gastrosteges, 170 to 185.

l. Bluish black with squarish red spots on the blanks; abdomen red with black spots. . . . . FARANCIA, 14.

ll. Blue-black, with three red lines; abdomen yellowish with a series of dark spots. . . . . ABASTOR, 15.

1. HETERODON, Beauvais. SPREADING ADDERS.

1. ***H. platyrhinus,*** Latreille. BLOWING VIPER. HOG-

NOSE SNAKE. Brownish, with about 28 dark dorsal blotches, besides lateral ones and half rings on the tail; sometimes uniform black. Vertical plate longer than broad, about equal to occipitals; L. 30; G. 125 to 150; scales 25. E. U. S., abundant. A very variable species; when angry it depresses and expands the head, hissing furiously, thus exhibiting a very threatening appearance, but it is perfectly harmless.

2. ***H. simus,*** (L.) Holbrook. HOG-NOSED SNAKE. Dorsal blotches about 35; ground color usually paler or yellowish brown; vertical plate much longer than occipitals, broader than long; G. 130; scales 23 to 27, usually 25. Southern, N. to Ills. and Wis.

2. *TROPIDONOTUS,* Kuhl. WATER SNAKES.

* Scales in 23 to 29 rows. (*Nerodia*, B. & G.)

1. ***T. sipedon,*** (L.) Holbr. WATER SNAKE. WATER ADDER. Brownish; back and sides with each a series of large, square, dark blotches alternating with each other; rarely uniformly dusky;. scales 23; G. 130 to 150; L. 30 to 50. E. U. S., abundant; aquatic.

Var. ***erythrogaster,*** (Shaw) Cope. RED-BELLIED WATER SNAKE. Uniform red-black above; copper-color below; head elongated. Michigan to Kansas and S.

Var. ***woodhousei,*** (B. & G.) Cope. Scales in 25 rows. Mo. to Texas.

2. ***T. rhombifer,*** Hallowell. HOLBROOK'S WATER SNAKE. Brown, with black quadrangular blotches; scales in 27 rows. Mich., Ills. and S.

** Scales in 19 to 21 rows. (*Regina*, B. & G.)

3. ***T. rigidus,*** (Say) Holbr. STIFF SNAKE. Greenish brown; two brown dorsal bands; abdomen yellowish, spotted; outer row of scales smooth; scales 19; G. 130 to 170; L. 24. Penn. to Ga.

4. ***T. leberis,*** (L.) Holbr. LEATHER SNAKE. Chestnut brown; a yellow lateral band and three narrow black dorsal stripes; scales all carinated; scales 19; G. 140 to 150; L. 24. U. S., chiefly eastward.

5. ***T. grahami,*** (B. & G.) Cope. GRAHAM'S SNAKE. Brown; a broad yellowish lateral band; scales all strongly carinated; head slender; abdomen unspotted; scales 19 (to 21?); G. 160; L. 20. Mississippi Valley, N. to Michigan.

3. *TROPIDOCLONIUM,* Cope. LITTLE RED SNAKES.

1. ***T. kirtlandi,*** (Kenn.) Cope. KIRTLAND'S SNAKE. Head shiny black; vertical plate broad; scales 19, all carinated; G. 130 to 140; L. 8. Ohio to Ill.; a handsome little snake.

4. *STORERIA,* Baird and Girard. RED-BELLIED SNAKES.

= *Ischnognathus*, Dum. & Bibron.

1. ***S. occipitomaculata,*** (Storer) B. & G. RED-BELLIED SNAKE. Grayish or chestnut brown, usually showing a paler vertebral band bordered by blackish dots; obscure dots on side; occiput with three pale blotches (very constant); belly salmon red; scales 15; G. 120 to 125; L. 12. U. S., chiefly eastward; abundant.

2. ***S. dekayi,*** (Holbr.) B. & G. DEKAY'S BROWN SNAKE. Grayish brown; a clay-colored dorsal band, bordered by dotted lines; grayish below; body thickish, tapering towards the small head; scales 17; G. 125 to 130; L. 12. E. U. S.

5. *EUTÆNIA,* Baird and Girard. GARTER SNAKES.

* Body very slender, elongated; tail nearly one-third of total length; scales in 19 rows.

1. ***E. saurita,*** (L.) B. & G. RIBAND SNAKE. SWIFT

GARTER SNAKE. Brown with three yellow stripes; light, clear brown, below the lateral stripes; tail usually more than $\frac{1}{3}$ of length; colors bright; G. 150 to 160; L. 36. U. S., chiefly E. of the Alleghanies.

2. ***E. faireyi,*** B. & G. FAIRIE'S GARTER SNAKE. Blackish, with three greenish yellow stripes; body relatively stout; tail less than $\frac{1}{3}$ length; space below bands same color as above; G. 165 to 180; L. 30. Miss. Valley, N. to Wis.

3. ***E. proxima,*** (Say) B. & G. SAY'S GARTER SNAKE. Blackish, dorsal stripe brownish yellow; lateral stripes greenish; tail $\frac{2}{7}$ of total length; sides colored like back; G. 165 to 175; L. 35. Miss. Valley, N. to Wis.

** Body stouter; tail shorter, about $\frac{1}{4}$ of total length; scales 19.

4. ***E. radix,*** Baird & Girard. HOY'S GARTER SNAKE. Black with three narrow yellow lines; scales very rough, the outer row broad; colors deep; head short; G. 150 to 160; L. 25. L. Michigan to Oregon.

5. ***E. sirtalis,*** (L.) B. & G. COMMON GARTER SNAKE. STRIPED SNAKE. Olivaceous, dorsal stripe narrow; lateral stripes rather broad but not conspicuous; colors generally duller than in the other species, lateral rows of spots more or less distinct; G. 130 to 160. N. Am., every where; our commonest snake; very variable. Prominent varieties are:

Var. ***ordinata,*** (L.) Cope, has the stripes duller and the spots more distinct, 85 in number. Chiefly northeastward.

Var. ***dorsalis,*** (B. & G.) Cope, has the dorsal stripe broad, and a row of distinct spots above the lateral stripe. N. Am., every where.

Var. ***parietalis,*** (Say.) Cope, has the stripes dull

greenish and the spaces between the lateral spots vivid brick red. Ind. (*Jordan*) and W.

6. BASCANIUM, Baird and Girard. BLACK SNAKES.

> *Coryphodon*, Dumeril et Bibron.

1. ***B. constrictor,*** (L.) B. & G. BLACK SNAKE. Lustrous pitch black, greenish below, chin and throat white; young olive with rhomboid blotches; large, rather slender; scales 17 (rarely 19); G. 170 to 200; L. 50 to 60. E. U. S., common E. and S.

7. PITYOPHIS, Holbrook. PINE SNAKES.

1. ***P. melanoleucus,*** (Daud.) Holb. PINE SNAKE. BULL SNAKE. White, with chestnut brown blotches which are margined with black, besides other markings; scales 29; G. 220 to 230; L. 60. Pine woods; N. J. to Ohio and southward.

2. ***P. sayi,*** (Schlegel) B. & G. WESTERN PINE SNAKE. Whitish or reddish, with many dark blotches and spots; scales usually 25; G. 220 to 230; L. 40 to 70. Western, E. to Wis

8. COLUBER, Linnæus. RACERS.

> *Scotophis*, B. & G.

1. ***C. guttatus,*** L. SPOTTED RACER. Red brown with a dorsal series of large, red, dark-edged blotches; belly checkered with black; scales 27; G. 210 to 230; L. 50. Virginia and S.

2. ***C. obsoletus,*** Say. PILOT SNAKE. RACER. Lustrous black, some scales white-edged; vertical plate longer than broad; scales 27; G. 235; L. 50 to 75. Mass. to Ills. and Texas; one of our largest snakes. (*C. alleghaniensis*, Holbr.)

3. ***C. vulpinus,*** (B. & G.) Cope. FOX SNAKE. Light brown, with quadrate, chocolate-colored blotches; vertical plate broader than long; scales 25; G. 200 to 210; L. 60. Mass. to Kansas and Northward.

4. ***C. emoryi,*** (B. & G.) Cope. EMORY'S RACER. Ashy gray with transverse brown blotches; vertical plate elongated; 6 or 8 median rows of scales only carinated; scales 29; G. 210 to 220; L. 40 to 50. Ills. to Kansas and Texas.

9. *CYCLOPHIS,* Günther. SUMMER SNAKES.

Leptophis, B. & G.

1. ***C. æstivus,*** (L.) Günther. SUMMER GREEN SNAKE. Head conical, neck very small; bright clear green, yellowish below; scales 17; G. 150 to 160; L. 30. Southern, N. to N. J. and Ills., abundant in the mountains; a most exquisite little creature.

10. *LIOPELTIS,* Fitzinger. GREEN SNAKES.

1. ***L. vernalis,*** (DeKay) Jan. GREEN SNAKE. GRASS SNAKE. SPRING SNAKE. Head elongate, neck slender; eyes very large; uniform deep green (bluish in spirits), yellowish below; scales 15; G. 130 to 140; L. 20. E. U. S., chiefly northerly; a beautiful species.

11. *DIADOPHIS,* Baird and Girard. RING-NECKED SNAKES.

< *Ablabes*, D. & B.

1. ***D. punctatus,*** (L.) B. & G. RING-NECKED SNAKE. Head depressed; eye rather large; blue-black above, bright pale orange below (yellowish in spirits); each plate usually with a black spot; yellowish occipital ring conspicuous; scales 15; G. 140 to 160; L. 15. Whole U. S.

Var. ***amabilis,*** (B. & G.) Cope, is slender, with 180 or more gastrosteges; below darker and more spotted. Western, E. to Ohio.

2. ***D. arnyi,*** Kenn. ARNY'S RING-NECKED SNAKE. Lead black; belly spotted and mottled with black; occipital ring narrow, scales 17. Ills. to Kansas.

12. ***OPHIBOLUS,*** Baird and Girard. KING SNAKES.

< *Coronella*, Laurenti.

< *Lampropeltis*, Auct.

1. ***O. getulus,*** (L.) B. & G. CHAIN SNAKE. THUNDER SNAKE. Black with narrow yellowish lines forking on the flanks, each fork embracing a large black spot; belly checkered; scales 21; G. 210 to 225; L. 50. Maryland to La., E. of the mountains; variable; represented westward by

Var. ***sayi,*** (Holbr.) Cope. KING SNAKE. Lustrous black, many scales with a whitish spot in the center. Alleghany to Rocky Mts., abundant, N. to Ills.; a handsome species.

2. ***O. doliatus,*** (L.) B. & G. RED SNAKE. CORN SNAKE, etc. Red with twenty to twenty-five pairs of black rings, each set enclosing a yellowish one; head red; scales 21; G. 180 to 210; L. 30 to 50. Md. to Kansas and S.; exceedingly variable, running by degrees into the following variety, extremes of which bear little resemblance to the typical *doliatus.*

Var. ***triangulus,*** (Boie.) Cope. MILK SNAKE. HOUSE SNAKE. SPOTTED ADDER. Grayish, with three series of brown, rounded blotches bordered with black, about fifty of them in the dorsal row; an arrow-shaped occipital spot; scales, etc., as in preceding. Va. to Iowa, and northward; very common. [*Coronella eximia*, (DeKay) Jan.]

3. ***O. calligaster,*** (Say.) Cope. Kennicott's Chain Snake. Light olive gray, with about sixty quadrate, chestnut colored, emarginate blotches on back and two rows of smaller ones on each side; scales in 25 rows. Ills. to Kansas. (*O. evansi*, Kenn.)

13. OSCEOLA, Baird and Girard. Scarlet Snakes.

1. ***O. elapsoidea,*** (Holbr.) B. & G. Scarlet Snake. Brilliant red, with about fifteen pairs of jet black rings on body and three on tail, each pair enclosing a white ring; the black rings tapering towards the sides, the white ones spreading; resembles closely *O. doliatus*; scales 19; G. 175 to 180; L. 20. Va. to Fla.

14. FARANCIA, Gray. Horn Snakes.

< *Calopisma*, D. & B.

1. ***F. abacura,*** (Holb.) B. & G. Red-Bellied Horn Snake. Blue-black with red lateral spots; eyes small; scales 19; G. 175; L. 36. Southern, N. to Ills. (*Nelson.*)

15. ABASTOR, Gray. Red-Sided Snakes.

1. ***A. erythrogrammus,*** (Daudin) Gray. Red-Lined Snake. Blue-black; sides with three red lines; eyes very large; nostrils in the middle of nasal plate; scales 19; G. 180; L. 25. Southern, N. to Ills. (*Nelson.*)

16. HALDEA, Baird & Girard. Brown Snakes.

= *Conocephalus*, D. & B.

1. ***H. striatula,*** (L.) B. & G. Brown Snake. Head

elongated, on a small neck; reddish gray, salmon red beneath; scales 17; G. 125 to 130; L. 10. Va. to Texas.

17. VIRGINIA, Baird and Girard. BROWN SNAKES.

1. ***V. valeriæ,*** B. & G. VALERIA BLANEY'S SNAKE. Head elliptical; body slender; brownish with minute black dots, often in two rows; yellowish beneath; scales 15; G. 120 to 130; L. 12. Md. to Ills. and S.

2. ***V. elegans,*** Kenn. KENNICOTT'S BROWN SNAKE. Scales much narrower; uniform olivaceous above; yellowish beneath; scales 17. S. Ills. to Ark.

18. CARPHOPHIOPS, Gervais. WORM SNAKES.
= *Celuta*, B. & G.

1. ***C. amœnus,*** (Say) Cope. GROUND SNAKE. Glossy chestnut brown; belly salmon-red; head very small; vertical plate broad; nasal plate large, pierced by the nostril; scales 13; G. 120 to 130; L. 12. Mass. to Ills. and S.

2. ***C. helenæ,*** (Kenn.) Cope. HELEN TENNISON'S SNAKE. Lustrous chestnut-brown, flesh color beneath; snout short and narrow; a single pair of frontal plates; scales 13. S. Ills. to Miss.

3. ***C. vermis,*** (Kenn.) Cope. WORM SNAKE. Purplish-black, two pairs of frontals, as in *C. amœnus;* belly flesh color, color extending on sides; scales 13; larger than the others. Missouri to Kansas.

FAMILY LXXV. (*b.*)—ELAPIDÆ.

(*The Harlequin Snakes.*)

Venomous snakes, provided with two or more permanently erect, grooved fangs in the upper jaw, and usually a series of smaller teeth behind them; scales not carinated; head usually quadrangular, with flat crown and short muzzle; no loral plate. Genera 15; species about 50, chiefly East Indian, a few inhabiting the warmer parts of America.

* Anal plate entire; urosteges two-rowed; scales in 15 rows. ELAPS, 1.

1. ELAPS, Schneider. HARLEQUIN SNAKES.

1. ***E. fulvius,*** (L.) Cuv. BEAD SNAKE. Jet black, with about 17 broad crimson rings, each bordered with yellow, and spotted below with black; a yellow occipital band; tail with yellow rings; L. 30; G. 200 to 215; U. 32. Va. to Ark. and S. A beautiful snake, mild in disposition and apparently harmless, although provided with venom-fangs. Resembles *Osceola* and *Ophibolus.*

FAMILY LXXVI.—CROTALIDÆ.

(The Crotalid Snakes.)

Upper jaw destitute of solid teeth, but provided with an erectile, grooved poison-fang on each side in front; a deep pit between eye and nostril. Tail often provided with a rattle, composed of horny rings of modified epidermis. Urosteges generally undivided, at least anteriorly. Scales carinated in all our species. Species 50, more or less, all American, and renowned for the deadliness of their venom.

* Tail with a rattle.

† Top of head covered with small, scale-like plates, larger ones in front; size large; rattle large. CROTALUS, 1.

†† Top of head with 9 large plates; size small; rattle small. CAUDISONA, 2.

** Tail without a rattle; general color chestnut, variegated. ANCISTRODON, 3.

1. CROTALUS, Linnæus. RATTLESNAKES.

1. ***C. horridus,*** L. BANDED OR NORTHERN RATTLESNAKE. Sulphur brown of various shades, with two rows

of confluent, brown, lozenge-shaped spots; tail black; a light loral line with a dark patch beneath; scales 23 to 25; G. 165 to 170; L. 40 to 60. U. S., chiefly eastward and southward, in rocky places; rapidly becoming extinct; several other species occur southwestward. (*C. durissus*, Auct.)

2. CAUDISONA, Laurenti. MASSASSAUGAS.

= *Crotalophorus*, Gray.

1. ***C. tergemina,*** (Say.) Cope. MASSASSAUGA. PRAIRIE RATTLESNAKE. Brown or blackish with about seven series of deep chestnut blotches, sometimes entirely black; scales 25; G. 140 to 150; L. 30. Prairie region, E. to the Alleghanies; abundant in grassy fields where not exterminated.

3. ANCISTRODON, Beauvais. COPPERHEADS.

> *Toxicophis*, Troost.

< *Trigonocephalus*, Holbr., etc.

1. ***A. contortrix,*** (L.) B. & G. COPPERHEAD. COTTON MOUTH. Hazel brown; top of head bright coppery, back with a series of fifteen to twenty-five dark blotches having something the form of an inverted **Y**; yellowish below with dark blotches; scales 23; G. 150 to 160; L. 35 to 40. E. U. S., chiefly southerly.

2. ***A. atrofuscus,*** (Troost) B. & G. BLACK MOCCASIN. Dusky above, with smoky gray blotches; tail black; belly white, blotched with black and minutely punctate; upper lip white; scales 25; G. 130 to 140; L. 25. Mts. of Tenn. and N. C.

3. ***A. piscivorus,*** (Holbr.) Cope. WATER MOCCASIN. Greenish brown with dark vertical bars; scales 25; G. 140; L. 30. Aquatic; southern, probably not in our limits.

Class IV.—Batrachia.

(*The Batrachians.*)

Cold-blooded Vertebrates, allied to the fishes, but differing in several respects, notably in the absence of rayed fins, the limbs being usually developed and functional, with the skeletal elements of the limbs of Reptiles; toes usually without claws.

The Batrachians undergo a more or less complete metamorphosis; the young ("tadpoles") being aquatic and fish-like, breathing by means of external gills or branchiæ; later in life, lungs are developed and (excepting in *Proteida*) the gills disappear. Skin naked and moist (rarely having imbedded scales) and used to some extent as an organ of respiration. Heart with two auricles and a single ventricle.

Reproduction by means of eggs which are of comparatively small size, without hard shell, developed in water or in moist situations.

The Batrachians differ more from the Fishes in appearance than in reality, and they are sometimes combined into one group (*Ichthyopsida*), the Birds and Reptiles constituting another (*Sauropsida*).

ORDERS OF BATRACHIA.

* Body short, depressed; tail wanting in the adult; four developed limbs, the posterior being much enlarged. . ANURA, W.

** Body lengthened, with a distinct tail; hind limbs if present not specially elongated.

† With no external gills or branchiæ when adult; eyelids present. . . . . URODELA, X.

†† External branchiæ and gill clefts persistent through life; no eyelids. . . . . PROTEIDA, Y.

W. FAMILIES OF ANURA.

* Fingers and toes tapering or cylindrical, not dilated into a disk at their tips.

† With teeth in the jaws; toes completely webbed.

a. No spur at the heel; fingers often webbed; chiefly aquatic. RANIDÆ, 77.

aa. One of the bones of the heel forming a sharp, flat-edged spur; fingers scarcely webbed; chiefly terrestrial. SCAPHIOPIDÆ, 78.

†† Jaws toothless; toes webbed; skin more or less warty; terrestrial. . . . . BUFONIDÆ, 80.

** Fingers and toes dilated at their tips, forming a viscous disk; arboreal. . . . . HYLIDÆ, 79.

X. FAMILIES OF URODELA.

* No spiracles or openings in the sides of the neck in the adult.

† Sides (in our species) with a row of round red or yellowish spots which are bordered with black (these very rarely obsolete); belly dotted with black; tarsus and carpus ossified; vertebræ opisthocœlian (not biconcave). PLEURODELIDÆ, 81.

†† Spots, if any, not as above.

‡ Carpus and tarsus cartilaginous; tongue much smaller and more extensively free than in the next.

a. Vertebræ concave behind only (opisthocœlian.) DESMOGNATHIDÆ, 82.

aa. Vertebræ biconcave (amphicœlian.) PLETHODONTIDÆ, 83.

‡‡ Carpus and tarsus ossified; vertebræ amphicœlian; tongue large, thick, papillose, attached by its base, with a narrow free margin; salamanders usually of large size and dark colors. . . . . AMBLYSTOMIDÆ, 84.

** With a spiracle or rounded opening in each side of the neck: size large.

b. Limbs well developed; toes 4–5. . . . MENOPOMIDÆ, 85.

bb. Limbs rudimentary; toes 2–2 or 3–3. . AMPHIUMIDÆ, 86.

Y. FAMILIES OF PROTEIDA.

* Hind legs present; both jaws with teeth; form salamander-like. PROTEIDÆ, 87.

** Hind legs wanting; upper jaw toothless; form eel-like. SIRENIDÆ, 88.

ORDER W.—ANURA.

(The Tailless Batrachians.)

Body nearly or quite naked, short and broad; all four limbs present; tail wanting in the adult; young (tadpole) fish-like, with broad head, external branchiæ, a long tail, no limbs and no teeth; the intestinal canal very long, adapted for a vegetable diet; from this form by degrees it develops into the adult animal which is always more or less Frog-like.

FAMILY LXXVII.—RANIDÆ.

(The Frogs.)

Tailless Batrachians with the tongue adherent in front and more or less free behind; fingers four, toes five, both commonly webbed; ear well developed, jaws and usually vomer, with teeth; chiefly aquatic. Genera fourteen, species about fifty; very abundant in tropical America.

* Vomerine teeth present; no finger opposed to the others; tongue nicked behind; toes full-webbed. . . RANA, 1.

1. *RANA,* Linnæus. FROGS.

* Back with large distinct spots arranged in more or less regular rows; back with two conspicuous yellowish folds.

1. ***R. halecina,*** Kalm. LEOPARD FROG. COMMON FROG. General color greenish, often bright, sometimes brassy, with many pale-edged dark spots which lie in two irregular rows on back; usually two large spots

between eyes; legs barred above; belly pearly or yellowish, each side of back with a well-marked fold. N. Am., the commonest species.

2. ***R. palustris,*** Le Conte. PICKEREL FROG. Brownish with the spots square, in four rows; young golden green; body with two glandular folds on each side; slender. E. U. S.

** Back with small dark spots or none.

3. ***R. clamitans,*** Merrem. GREEN FROG. SPRING FROG. Bright green, darker on the flanks, every where spotted with blackish; color sometimes rather brown than green; white below; glandular folds conspicuous; size moderate. U. S., E. of the mountains. (*R. horiconensis*, Holbr. *R. fontinalis*, Le C.)

4. ***R. catesbiana,*** Shaw. BULL FROG. Greenish, of varying shades, with numerous small, indistinct darker spots, head usually of a very bright pale green; glandular folds little marked; very large, a foot or more long when adult. E. U. S., common; well noted for its rich bass notes. (*R. pipiens*, Auct., not of L.)

5. ***R. temporaria,*** (L.) var. ***sylvatica,*** (Le C.) Gthr. WOOD FROG. Color reddish brown; a dark band on each side of head through eye and ear; quite small. E. U. S., common; scarcely aquatic. (*R. cantabrigensis*, Baird, the Cambridge Frog, from E. Mass., N. and W. is another variety of this European species.)

FAMILY LXXVIII.—SCAPHIOPIDÆ.

(*The Spade Foots.*)

Terrestrial frogs having the heel provided with a more or less developed spur. Genera three; species ten or more. In Europe, America, and Australia.

* Toes completely webbed; forehead and crown bony, rough; skin slightly tuberculate. . . . SCAPHIOPUS, 1.

1. SCAPHIOPUS, Holbrook. Spade Foots.

1. ***S. holbrookii,*** (Harlan) Baird. Solitary Spade Foot. Heel with a sharp-edged spur; olive brown, a pale yellow streak on each side. E. U. S., not very common; burrows in the ground. (*S. solitarius*, Holbr.)

FAMILY LXXIX.—HYLIDÆ.

(*The Tree Frogs.*)

Arboreal frogs of small size, having the fingers and toes more or less dilated into disks at their tips; ear well developed. Genera ten; species sixty; found in most parts of the world; noted for their shrill voices.

* Disks round, conspicuous; fingers somewhat webbed; skin roughened. . . . . Hyla, 1.

** Disks small; fingers not webbed.

† Toes webbed only at base or not at all; tympanum distinct. Chorophilus, 2.

†† Toes broadly webbed; tympanum indistinct. . Acris, 3.

1. HYLA, Laurenti. Tree Frogs.

1. ***H. versicolor,*** LeConte. Common Tree Toad. Green, gray or brown, with irregular dark spots; below yellow or white; fingers one-third webbed; exceedingly variable. E. U. S., very abundant.

2. ***H. pickeringii,*** Holbrook. Pickering's Tree Toad. Yellowish brown with dusky rhomboidal spots and lines sometimes arranged in the form of a cross. E. U. S.

3. ***H. andersonii,*** Baird. Anderson's Tree Toad. Deep pea-green; sides with irregular yellow spots; a purplish band on sides of head. N. J. to S. C., rare

2. CHOROPHILUS, Baird. Little Tree Frogs.

1. ***C. triseriatus,*** (Wied.) Baird. Tree Frog. E. U. S.

3. ACRIS, Dumeril and Bibron. CRICKET FROGS.

1. ***A. gryllus,*** (Le C.) var. ***crepitans,*** (Baird) Cope. CRICKET FROG. Brownish above; middle of back and head bright green; a dark triangle between the eyes; sides with three oblique blotches; a white line from eye to ear. E. U. S. (the typical *gryllus* southward.)

FAMILY LXXX.—BUFONIDÆ.

(*The Toads.*)

Maxillaries toothless; toes webbed, not dilated at their tips; ear well developed; skin usually warty. Genera three, species thirty; in every part of the world except Australia. Most of them belong to the familiar genus, *Bufo.*

1. BUFO, Laurenti. TOADS.

1. ***B. lentiginosus,*** Shaw. AMERICAN TOAD. Brownish olive with a yellowish vertebral line and some brownish spots; adults very warty; young nearly smooth. U. S., very common, variable; the northern form is var. *americanus* (Le C.) Cope.

ORDER X.—URODELA.

(*The Salamanders.*)

Body naked, elongated, subcylindrical; four limbs developed; tail persistent, usually much longer than broad, terete or compressed; no external branchiæ when adult.

FAMILY LXXXI.—PLEURODELIDÆ.

(*The Newts.*)

Vertebræ concave behind only (opisthocœlian); carpus and tarsus ossified. I am unable to characterize this family further.

* Tongue small, thick, oval, attached by nearly its whole inferior surface; toes 4–5, outer and interior on hind foot rudimentary; our species spotted. . . . . DIEMYCTYLUS, 1.

1. DIEMYCTYLUS, Rafinesque. SPOTTED NEWTS.

> *Notophthalmus*, Raf.

1. ***D. viridescens,*** Raf. SPOTTED TRITON. NEWT. EVET. EFT. Above olive green of varying shades; lemon yellow below; each side with a row of several rather large vermillion spots, each surrounded by a black ring; back with a pale streak; belly, etc., with small black dots. E. U. S., abundant E. of the Alleganies; in ponds and brooks.

2. ***D. miniatus,*** Raf. RED EFT. RED EVET. Color vermillion red of varying shade, paler or yellowish below; markings precisely as in the foregoing; same range, but found away from water, under stones, etc.; comes out after rain. Perhaps a variety or state of the preceding, at least Prof. Cope so considers it.

FAMILY LXXXII.—DESMOGNATHIDÆ.

(*The Desmognaths.*)

Vertebræ opisthocœlian; carpus and tarsus cartilaginous; no crests or other dermal appendages developed at the breeding season. Genus one; species three; all of the Eastern U. S. In external characters, this family is scarcely distinguishable from the next, but the skeletal distinctions are quite numerous. They are, however, too technical for our present purpose. "The examination of the skeleton of species of this genus utterly changes the impressions produced by a consideration of the external characters. It may be stated as characteristic of the Batrachia in general, that their affinities can not be determined without study of the skeleton." *Cope.* Proc. Phil. Ac. Nat. Sc. 1869, 113.

1. DESMOGNATHUS, Baird. DUSKY SALAMANDERS.

1. ***D. ochrophæa,*** Cope. YELLOW DESMOGNATH. Brownish yellow with a brown shade on each side; a yellowish dorsal band; back with a few spots; belly unspotted; tail rounded; ♂ with lower jaw toothless behind; costal folds 14; size small; scarcely aquatic. Allegany Mountains, N. Y., southward.

2. ***D. fusca,*** (Raf.) Baird. DUSKY SALAMANDER. Brown above, with gray or purplish spots or shades, becoming blackish with age; marbled below; eyes prominent; tail compressed and keeled, as long as head and body; costal folds 14; larger. Mass. to Ohio and S.; one of the commonest species in springs and brooks; remarkable for its activity and strength.

3. ***D. nigra,*** (Green) Baird. BLACK SALAMANDER. Uniform black; tips of tail, jaws, etc., brown; tail compressed and finned; costal folds 12. Penn., S. in the mountain springs; the largest Eastern Salamander.

FAMILY LXXXIII.—PLETHODONTIDÆ.

(*The American Salamanders.*)

Vertebræ amphicœlian; carpus and tarsus cartilaginous. Various other distinctive characters are given by Prof. Cope, but we omit them here as not available for our present purpose. Genera eleven; species thirty; nearly all North American.

* The tongue attached by a band running from its central or posterior pedicel to the anterior margin; premaxillaries 2.

† Toes 4–4, small, ashy above, spotted below. HEMIDACTYLIUM, 1.

†† Toes 4–5; colors dark; spotted or banded. PLETHODON, 2.

** Tongue free all around, attached by its central pedicel only; toes 4–5, all free.

‡ Premaxillaries united; color yellow or red, spotted or striped. SPELERPES, 3.

‡‡ Premaxillaries 2; color purplish gray or salmon color, unspotted. . . . . GYRINOPHILUS, 4.

1. *HEMIDACTYLIUM*, Tschudi. FOUR-TOED SALAMANDERS.

1. ***H. scutatum***, (Schl.) Tsch. FOUR-TOED SALAMANDER. Ashy brown above; snout yellow; silvery below, with dots like ink spots; tail slender, nearly twice the length of the body; head blunt. R. I. to Ills., and S. (S. *melanosticta*, Gibbes.)

2. *PLETHODON*, Tschudi. PLETHODONTS.

1. ***P. erythronotus***, (Green) Baird. RED-BACKED SALAMANDER. Plumbeous above, often with a broad red dorsal band; belly marbled; body very slender; tail cylindric; inner toes rudimentary; costal folds 16 to 19. E. U. S., common. [*P. cinereus*, (Green) Cope, variety without red dorsal band.]

2. ***P. glutinosus***, (Green) Baird. VISCID SALAMANDER. Black, usually with gray lateral blotches and smaller dorsal spots; stout; tail rounded; inner toes well developed. E. U. S., chiefly terrestrial, like the preceding.

3. *SPELERPES*, Rafinesque. CAVE SALAMANDERS.

1. ***S. bilineatus***, (Green) Baird. TWO-STRIPED SALAMANDER. Yellow with a dark line along each side of the back; belly unspotted; tail not keeled; costal folds 14; small. Maine to Wis. and S.

2. ***S. longicaudus***, (Green) Baird. CAVE SALAMANDER. Lemon yellow; sides with many small black spots; a median dorsal series; belly spotless; tail keeled, very

long, spotted or barred with black; costal folds 13; large. Maine to Minn. and S., abounding in the caves of Ky. and Ind.

3. ***S. ruber,*** (Daudin) Gray. RED TRITON. Vermilion red, with black or brown spots; head wide; costal folds 15 or 16; large; variable. Maine to Nebraska and S. (*Pseudotriton*, Baird.)

4. ***GYRINOPHILUS,*** Cope. PURPLE SALAMANDERS.

1. ***G. porphyriticus,*** (Green) Cope. PURPLE SALAMANDER. Uniform purplish gray above; head broad; tail rounded at base, not finned; large; aquatic. Allegany Mountains, N. E. and S. [*Spelerpes salmonea* (Stor.) Gray.] "The only one of our Eastern Salamanders which attempts self defense. It snaps fiercely but harmlessly and throws its body into contortions in terror." (*Cope.*)

FAMILY LXXXIV.—AMBLYSTOMIDÆ.

(*The Amblystomas.*)

Vertebræ amphicœlian; carpus and tarsus ossified; digits 4–5, without webbing; tongue thick; size generally large and color dark. Genus one, species nineteen; probably all American, and very abundant in the Southern and Western parts of the U. S. The larvæ which reach a large size, and even breed before the gills are absorbed, have long been considered as forming a separate genus, *Siredon*, supposed to be allied to *Necturus*.

1. ***AMBLYSTOMA,*** Tschudi. BIG SALAMANDERS.

* Folds of tongue radiating from behind; costal folds 10 to 12; fourth toe with 4 phalanges.

† Costal grooves 10.

1. ***A. talpoideum,*** (Holbr.) Gray. MOLE SALAMANDER.

Blackish brown, gray-speckled; tail short, compressed, $2\frac{1}{2}$ in length; head very broad; body short and squat. Southern, N. to S. Ills.

†† Costal grooves 11.

‡ Sole with one indistinct tubercle, or none.

2. ***A. opacum,*** (Gravenhorst) Baird. OPAQUE SALAMANDER. Black above, with bluish gray bars; belly dark blue; no dorsal furrow, no enlarged pores on the head; tail $2\frac{1}{2}$ in total length; body stout. Penn. to Wis. and S. A handsome species. (*S. fasciata*, Green.)

3. ***A. punctatum,*** (L.) Baird. LARGE SPOTTED SALAMANDER. Black above with a series of round yellow spots on each side of the back; body broad, depressed and swollen; skin punctate with small pores, from which exudes a milky fluid (*Cope*); two or three clusters of enlarged pores on head; a strong dorsal groove; tail $2\frac{1}{3}$ in length; large. U. S., E. of the Rocky Mountains. (*A. venenosa* and *subviolacea*, Auct.)

4. ***A. conspersum,*** Cope. SMALLER SPOTTED SALAMANDER. Lead colored, with one or two series of small yellow spots along sides; no dorsal groove; skin smooth; tail $2\frac{1}{2}$ in length; small. Penn. to Ga.

‡‡ Sole with two distinct tubercles.

5. ***A. bicolor,*** Hallowell. TWO-COLORED SALAMANDER. Olive brown, yellowish below, rising in blotches on the sides; a few large yellowish spots above; limbs banded; tail yellow with brown spots. New Jersey.

††† Costal grooves 12.

a. Large species; sole with two distinct tubercles.

6. ***A. tigrinum,*** (Green) Baird. TIGER SALAMANDER. Chiefly brown with many yellow spots, about as large as the eye; body thick and strong; the head comparatively long and narrow; tail shorter than head and body; color

varying from uniform brown to yellow, but usually spotted. U. S., E. of the Rocky Mountains. (*A. ingens*, Hallowell.)

7. ***A. xiphias,*** Cope. LONG-TAILED SALAMANDER. Yellow-olive with brown reticulating bands; head small, blunt; tail very long, much longer than the head and body. Ohio.

aa. Small species; sole with one indistinct tubercle or none.

8. ***A. jeffersonianum,*** (Green) Baird. JEFFERSON'S SALAMANDER. Olive brown or blackish, usually with pale or bluish spots, but sometimes uniform plumbeous. Va. to Ind. and N., variable; several varieties are recognized by Prof. Cope.

** Folds on tongue radiating from the median longitudinal furrow; costal folds 12; size small.

9. ***A. microstomum,*** Cope. SMALL-MOUTHED SALAMANDER. Lead-colored, usually with gray shades and specks; head small, short, broad; body slender; skin very smooth and slippery; lower jaw projecting. Ohio to Kansas and S.

FAMILY LXXXV.—MENOPOMIDÆ.

(*The Menopomes.*)

Salamanders of large size, having the form of *Amblystoma*, but with an orifice on each side of neck persistent during life; no external gills; legs well developed; toes 4–5; aquatic. Genus one; species two, *Menopoma fuscum*, Holbr., of the head waters of the Tennessee, and the following:

1. MENOPOMA, Harlan. HELLBENDERS.

1. ***M. alleghaniense,*** Harlan. HELLBENDER. BIG WATER LIZARD. Blackish; length 1½ to 2 feet. Mississippi Valley to N. C. and S

FAMILY LXXXVI.—AMPHIUMIDÆ.

(*The Congo Snakes.*)

Salamanders of large size, having the body elongated almost serpentiform; limbs rudimentary with two or three toes each; a spiracle on each side of neck as in the preceding family; aquatic. Genera two; the three-toed *Murænopsis* (*M. tridactylus*) and the two-toed *Amphiuma*. Species two, inhabiting the ditches and streams of the warmer parts of the U. S.

1. AMPHIUMA, Linnæus. TWO-TOED CONGO SNAKES.

1. ***A. means,*** L. CONGO SNAKE. Dusky; limbs very small, each with two toes. Southern States, N. to N. C.

ORDER Y.—PROTEIDA.

(*The Proteans.*)

Tailed Batrachians, provided with bushy external gills which are persistent during life; lungs more or less developed and functional, hence these animals are truly amphibious.

FAMILY LXXXVII.—PROTEIDÆ.

(*The Mud Puppies.*)

Salamanders of medium or large size, provided with bushy external gills, and having the branchial clefts remaining open through life; teeth well developed. Genera two. *Proteus* inhabitating caves in S. W. Austria (*Carniola*), and *Necturus* of the fresh waters of the U. S. *Proteus* is blind and has the toes 3–2. *Necturus* has the eyes well developed, though small, and the toes 4–4.

1. ***NECTURUS,*** Rafinesque. MUD PUPPIES.

= *Menobranchus*, Harlan.

1. ***N. lateralis,*** (Say.) Baird. MENOBRANCHUS. MUD PUPPY (North). WATER DOG (South). DOG FISH. Brown, more or less spotted; young with traces of a lateral band; dusky below; gills large and bushy, bright red, forming three tufts on each side; head broad, depressed; tail much compressed. E. U. S., chiefly northern and west of the Alleganies, especially abundant in the Great Lake Region; reaches a length of eight inches to two feet. (*M. maculatus*, *hyemalis*, etc., of authors.) Another species. *N. punctatus*, (Gibbes) Cope, occurs in S. C.

FAMILY LXXXVIII.—SIRENIDÆ.

(*The Sirens.*)

Body elongated, eel-like; external branchiæ persistent; no posterior limbs, not even a vestige of pelvis; head flattened; snout obtuse; mouth narrow, the lower jaw with teeth all around, the upper toothless; eye very small. (*Cuvier.*) Genera two, species two,—*Pseudobranchus striatus*, (LeC.) of Georgia, a small species with small gills, and the following:

1. ***SIREN,*** Linnæus. SIRENS.

1. ***S. lacertina,*** L. GREAT SIREN. Reaches a length of three feet. Southern, N. to N. C. and S. Ills.

Class V.—Pisces.

(*The Fishes.*)

A fish is a cold-blooded vertebrate, adapted for life in water, having the limbs developed as fins, the fingers and toes being represented by cartilaginous rays connected by membrane (in rare cases limbs rudimentary or wanting); exoskeleton usually developed as scales or bony plates (skin rarely naked); one or more fins on the median line of the body, composed of rays connected by membrane. Skull developed, containing a brain of several differentiated ganglia; a distinct lower jaw. Heart with an auricle, ventricle, and arterial bulb; respiration by means of branchiæ, which consist (typically) "of bony arches attached to the hyoid bone, to which the filaments of the gills are attached, generally in a row upon each, and having their surface covered by a tissue of innumerable blood vessels. The water taken in at the mouth passes among the filaments of the gills and escapes by the gill openings towards the rear; in its progress through the filaments of the gills the water imparts to these the oxygen of the air which it contains. The blood is sent to the gills by the heart, which thus answers to the right side of the heart of warm-blooded animals, and from the gills it is sent to an arterial trunk lying along the under side of the vertebral column, which distributes the blood through the body of the fish" (*Cuvier*); branchiæ free, gill openings a single cleft on each side. In most fishes there is a membranous air bladder immediately beneath the back-bone, answering homologically to the lungs of the higher vertebrates; in a few Ganoids the air bladder is cellular, and more or less functional and connected by a glottis with the œsophagus; in most of the soft-rayed *Teleocephali* there is a slender duct connecting the air bladder with the alimentary canal; in the *Acanthopteri* and others this is wanting. Reproduction by eggs of small size, which are fertilized generally after exclusion; a few are ovoviviparous.

As here characterized, the class *Pisces* includes the *Teliosts* and *Ganoids*, of authors, and excludes the Sharks and Skates and their allies, as well as the Lampreys and Lancelets, which differ from the true Fishes more than the latter do from the Batrachians.

The following key includes not only the families of fresh water fishes described in this work, but also all of the families of Fishes represented on the Atlantic Coast of the U. S. The names of those families which are exclusively marine are printed in italics, and no further reference is made to them. A student, therefore, who traces a fresh water fish to any of them will understand that there is an error on his part or mine. The key is, of course, purely artificial, and does not, in most cases, give true family distinctions, for instance:

With 5 to 9 detached finlets behind dorsal and anal; dorsals 2; scales small or none. . . *Scombridæ*, the *Mackerels*.

does not imply that all *Scombridæ* possess those characters, nor, indeed, that all possessing them are *Scombridæ;* but that all fishes in the region *here covered*, which show those peculiarities, are to be referred to that family.

SUB-CLASS I. Tail homocercal (caudal fin rarely wanting); optic nerves simply crossing, without chiasma; arterial bulb simple, with two opposite valves at its origin; air bladder, if present, not cellular; exoskeleton typically of scales, either ctenoid or cycloid. . . . . TELEOSTEI, page 201.

SUB-CLASS II. Tail heterocercal; optic nerves forming a chiasma; arterial bulb with several rows of valves; air bladder frequently cellular and lung-like; exoskeleton typically of bony plates. . . . . GANOIDEI, page 212.

ORDERS OF TELEOSTEI.

Gills pectinated—of the ordinary sort, not tuft-like.

I. Maxillaries normally developed and normally distinct from each other; gills not in the axils; typical fishes (characters too various to be here summarized). . TELEOCEPHALI, Z.

II. With 4 to 8 long barbels about the mouth, the longest of which is a continuation of the incomplete maxillary; subopercle wanting; ventrals abdominal; usually an adipose fin and dorsal and pectoral spines; skin naked or with bony plates; chiefly in fresh water.

NEMATOGNATHI, AA.

III. Maxillaries rudimentary or wanting; scapular arch free from skull; body elongated, serpentiform, with a long dorsal and anal, which meet around the tail; no ventral fins; scales small or none; jaws with teeth; chiefly marine. . . . . APODES, BB.

IV. Carpal bones elongated, forming a kind of arm which supports the pectorals, in the axils of which are the small gill openings; ventrals jugular, with 4 or 5 soft rays; body scaleless or tuberculate; head very large; marine.

PEDICULATI, page 211.

V. Intermaxillaries immovably united with the maxillaries; skin rough, often covered with spines or ganoid plates; ventral fins wanting; marine. PLECTOGNATHI, page 212.

** Gills small, tuft-like, largest at their free tips; body covered with bony plates; mouth small, toothless, at the end of the long snout; no ventral fins; marine.

LOPHOBRANCHII, page 212.

Z. SUB-ORDERS OF TELEOCEPHALI.

I. Body flat, unsymmetrical; both eyes on the upper or colored side; ventrals jugular. . . HETEROSOMATA, page 208.

II. Bones of snout prolonged into a long tube which bears the short jaws at the end. . . HEMIBRANCHII, page 209.

III. With two or more free spines in place of first dorsal; ventrals sub-abdominal, of a stout spine and a small ray; small fishes scaleless or with bony plates. . HEMIBRANCHII, page 209.

IV. Dorsal fins two, distinct, small, the first of 4 to 7 spines; ventrals abdominal; teeth feeble or wanting; scales cycloid, silvery. . . . . PERCESOCES, page 208.

V. With the first rays of the dorsal, or the whole first dorsal, of simple — usually stiff spines; first ray of ventral usually inarticulate (spinous dorsal forming a hump in *Cyclopterus*; a lamellated sucking disk in *Echeneis*, etc., wanting altogether in *Aspidophoroides* and *Gobiesox*). ACANTHOPTERI, page 203.

VI. Fin rays soft and articulated (excepting occasionally one or two in dorsal or anal); no ventral spines; scales when present, usually cycloid.

* Ventrals jugular; dorsal and anal long, often divided. ANACANTHINI, page 208.

** Ventrals abdominal.

† Mouth entirely toothless; abdomen not serrated; lower pharyngeals falciform, tooth-bearing; no adipose fin; head naked; fresh water. . EVENTOGNATHI, page 211.

†† Body elongated, scaly; a series of keeled scales along sides of abdomen; lower pharyngeals united (as in *Labridæ*); no air duct; no adipose fin; no ventral serratures; one or both jaws or else pectoral fins greatly elongated; chiefly marine. . . SYNENTOGNATHI, page 209.

††† Head more or less scaly (naked in *Amblyopsis*, the Cave Blind Fish); both jaws fully provided with teeth; lower jaw usually longest; dorsal far back, nearly opposite anal; no adipose fin, ventral serratures, nor peculiar scales; chiefly fresh water. . HAPLOMI, page 209.

†††† Soft-rayed fishes showing none of the above combinations of characters; head naked; adipose fin or abdominal serratures often present; dentition and habitat various. ISOSPONDYLI, page 210.

*** Ventrals entirely wanting.

‡ Jaws with teeth; vent at the throat; body oblong; cave fishes. . . . . . . HAPLOMI, page 209.

‡‡ Jaws toothless; vent normal; body serpentiform. ANACANTHINI, page 208.

FAMILIES OF ACANTHOPTERI.

1. With 5 to 9 detached finlets behind dorsal and anal; dorsals two; scales small or none. . *Scombridæ*, the *Mackerels*.

2. Upper jaw prolonged into a "sword"; teeth feeble or wanting; scaleless; size large. . . *Xiphiidæ*, the *Sword-Fishes*.

3. Tail ending in a sharp point; no caudal nor ventrals; teeth strong. . . . . *Trichiuridæ*, the *Hair-Tails*.

4. First dorsal on the top of head, modified into a lamellated sucking disk. . . . *Echeneididæ*, the *Remoras*.

5. Ventral fins completely united, sometimes forming a sucking disk.

   — Dorsals two, distinct; body scaly or not. . GOBIIDÆ, 96.

   — Dorsal single; spinous dorsal enveloped in skin, forming a hump in the adult; scaleless, tuberculate. *Cyclopteridæ*, the *Lump-Suckers*.

   — Dorsal single; body elongated; scaleless; small fishes often parasitic in shells of Mollusks. *Liparididæ*, the *Sea Snails*.

6. Ventral fins wide apart, with a sucking disk between them; dorsal spineless, on the tail. *Gobiesocidæ*, the *Pike-Suckers*.

7. With a stout, sharp spine on each side of tail; body much compressed. . . . . *Acanthuridæ*, the *Surgeons*.

8. With several unconnected spines in place of the first dorsal.

   — Tail with a keel on each side. *Carangidæ*, the *Pilot Fishes*.

   — Anal fin preceded by two free spines; body compressed and elevated. . . . *Carangidæ*, the *Pilot Fishes*.

   — Tail not keeled; jaws toothless; body very long and slender. *Ammodytidæ*, the *Sand Launces*.

   — Tail without a keel; jaws with teeth.

   Body long; snout elongated. *Elacatidæ*, the *Crab-Eaters*.

   Body short, compressed; snub-nosed. *Stromateidæ*, the *Harvest Fishes*.

9. With none of the preceding combinations.

   * With two distinct dorsal fins — rarely slightly connected by membrane at the base.

   † Body with developed scales or bony plates, large or small.

Small fresh water fishes (1 to 6 inches long); elongated or fusiform, often brightly colored; the fins — especially the pectorals—well developed; anal spines one or two; air bladder rudimentary. ETHEOSTOMIDÆ, 89.

2. Pectoral fins very long, reaching at least to anal, with 3 detached appendages or else several connected, forming an additional fin· cheeks mailed; head bony. *Triglidæ*, the *Gurnards*.

3. With 7 or 8 filiform appendages on each side below the pectorals; cheeks not mailed. *Polynemidæ*, the *Thread-Fishes*.

4. Throat with two long barbels. *Mullidæ*, the *Surmullets*.

5. Dorsal spines only two; scales minute, imbedded in the skin. . . . *Rhypticidæ*, the *Soap Fishes*.

6. Ventrals abdominal; body elongated; scales cycloid; teeth stout. . . *Sphyrænidæ*, the *Barracudas*.

7. With none of the above combinations; ventrals mostly thoracic.

a. Some or all of opercular bones, more or less serrated or spinous.

b. With teeth on the vomer.

c. First dorsal low and weak of 8 spines; scales small; one or more minute spines in front of anal; teeth strong. . *Pomatomidæ*, the *Blue Fishes*.

cc. Dorsal spines stout; scales ctenoid; no free anal spines.

d. Ventrals 1–5; branchiostegals usually 7.

e. Cleft of mouth horizontal or oblique; scales firm.

— Anal spines 2, sometimes obscure. PERCIDÆ, 90.

— Anal spines 3, distinct. . LABRACIDÆ, 91.

ee. Cleft of mouth nearly vertical; scales large, deciduous. *Chilodipteridæ*, the *Apogons*.

dd. Ventrals 1–7; branchiostegals 8; anal spines 4. *Berycidæ*, the *Berycoids*.

bb. No teeth on the vomer; anal spines 1 or 2; lateral line usually running up on the caudal fin.
SCIÆNIDÆ, 94.

aa. Edges of opercular bones entire.

f. Scales well developed, not enlarged along lateral line; chin often with barbels; no free spines.
Sciænidæ, the *Maigres*.

ff. Scales minute; no barbels.

g. Body more or less compressed and elevated; scales sometimes enlarged along lateral line; usually 2 free anal spines. *Carangidæ*, the *Pilot Fishes*.

gg. Body long and low; no free spines nor lateral shields. . . . . *Gobiidæ*, the *Gobies*.

†† Body entirely scaleless.

h. Body more or less depressed; eyes high up on the broad head; caudal usually rounded.

i. Dorsal with 4 spines; ventrals jugular I, 5; mouth vertical. . . *Uranoscopidæ*, the *Star Gazers*.

ii. Dorsal with 3 spines; ventrals jugular I, 2; mouth broad, with conical teeth.
Batrachidæ, the *Toad Fishes*.

iii. Dorsal spines 6 or more; ventrals thoracic; cheeks mailed (*i. e.*, the sub-orbital bone extending backward over the cheek, articulating with the preopercle).

— Spinous dorsal shortest, its middle rays highest; head without barbels. . . COTTIDÆ, 95.

— Spinous dorsal longest, notched, its first rays highest; head with many fleshy slips.
Hemitripteridæ, the *Sea Ravens*.

hh. Body greatly compressed; the eyes lateral or anterior; fins often filamentous; tail usually slender, the caudal fin widely forked.

j. A series of bony shields along base of second dorsal.
Zenidæ, the *John Dories*.

jj. No bony shields; usually two free anal spines.
Carangidæ, the *Pilot Fishes*

** Dorsal fin single, not divided to its base.

k. Fresh water species.

l. Ventrals I, 5; dorsal spines 8 to 12; vent normal.

m. Teeth on vomer; anal spines 3 to 9. ICHTHELIDÆ, 92.

mm. No teeth on vomer; anal with 2 (or 1) spines, the second very strong. . . . SCIÆNIDÆ, 94.

ll. Ventrals 7-rayed; dorsal spines 3; vent jugular. APHREDODERIDÆ, 93.

kk. Marine species.

1. Cheeks mailed (as in *Cottidæ*, etc.)

n. Body covered with bony, keeled plates; no dorsal spines. . . . *Agonidæ*, the *Sea Poachers*.

nn. Body with ordinary scales; spinous dorsal many-rayed. . . *Scorpænidæ*, the *Sea Scorpions*.

2. With broad, cutting, incisor-like front teeth, or with crushing, molar-like lateral teeth or both; scales rather large; usually a recumbent free spine in front of the dorsal fin.

o. Opercular bones entire; vertical fins not much scaly. *Sparidæ*, the *Sea Breams*.

oo. Preopercle denticulated; soft parts of vertical fins densely scaly. . *Pimelepteridæ*, the *Fat-Fins*.

3. Body much compressed and elevated; the soft rays of the vertical fins covered high up with ctenoid scales; teeth villiform; body often dark-banded.

p. Dorsal with less than 10 spines, separated by a notch from the soft part; spinous dorsal scaleless. *Ephippidæ*, the *Moon Fishes*.

pp. Dorsal undivided, with 10 or more spines, scaly throughout; fins often filamentous. *Chætodontidæ*, the *Chætodonts*.

4. Ventrals wanting; scales minute; body high, much compressed. . *Stromateidæ*, the *Harvest Fishes*.

5. Ventrals jugular, few-rayed or wanting; body long and low; dorsal fin very long, occupying most of the back, at least half of it and sometimes all composed of flexible spines; scales small or none; usually an anal papilla.

q. Ventrals present.

r. Dorsal with both spines and soft rays.
Blenniidæ, the *Blennies*.

rr. Dorsal composed of spines only.

s. Lateral line usually present and sometimes duplicated; head conic; compressed; pyloric cœca present. . *Stichæidæ*, the *Snake Blennies*.

ss. No lateral line; ventrals I, 1; no pyloric cœca; teeth, small, acute. *Xiphidiontidæ*, the *Gunnels*.

qq. No ventral fins.

t. Gill openings wide; scales rudimentary; cleft of mouth not vertical; teeth strong.
Anarrhichadidæ, the *Wolf Fishes*.

tt. Gill openings moderate; no scales; cleft of mouth nearly vertical; dorsal of spines only; body almost eel-like. . *Cryptacanthidæ*, the *Ghost Fishes*.

6. With none of the preceding combinations.

u. Ventrals jugular, 4-rayed; dorsal very high and long.
Bramidæ, the *Winged Dolphins*.

uu. Ventrals thoracic, I, 5.

v. Dorsal of 50 or more rays running from head to tail; the spinous part not differentiated.
Coryphænidæ, the *Dolphins*.

vv. Dorsal shorter, the two sorts of rays different.

w. Lateral line interrupted.

x. Scales ctenoid; dorsal spines 13; depth more than half length.
Pomacentridæ, the *Demoiselles*.

xx. Scales cycloid; dorsal spines 9; depth less than half length. . *Labridæ*, the *Wrasse Fishes*.

ww. Lateral line continuous.

y. Opercle or preopercle or both distinctly serrated.

z. Spinous dorsal longer than the soft part, of 18 spines which are tipped with little membranous appendages; scales cycloid.
Labridæ, the *Wrasse Fishes*.

zz. Dorsal spines 8, very low, nearly equal and scarcely connected; snout blunt.
Stromateidæ, the *Harvest Fishes.*

zzz. With neither of the preceding combinations.

a. No teeth on vomer.
Pristipomatidæ, the *Red Mouths.*

aa. Teeth on vomer.

b. Canines present; branchiostegals 7.
Serranidæ, the *Sea Bass.*

bb. No canines; branchiostegals 6; eyes large. . *Priacanthidæ*, the *Big Eyes.*

yy. Opercular bones with entire edges.

c. Dorsal with nine to 20 spines; anal III, 8 or more; lips large, fleshy.
Labridæ, the *Wrasse Fishes.*

cc. Dorsal with 9 spines; anal III, 7.
Gerridæ, the *Gerroids.*

FAMILIES OF ANACANTHINI.

* Ventrals jugular, sometimes rudimentary.

† Caudal fin developed as a separate fin; lateral line continuous.
GADIDÆ, 97.

†† Caudal fin not separate, dorsal and anal confluent around the tail.

‡ Ventral fins developed, 4-rayed. *Lycodidæ*, the *Eel Pouts.*

‡‡ Ventral fins replaced by a pair of bifid filaments.
Ophidiidæ, the *Cusk Eels.*

** Ventral fins entirely wanting.

a. Vent remote from the head. *Ammodytidæ*, the *Sand Launces.*

aa. Vent at the throat; size small. *Fierasferidæ*, the *Fierasfers.*

FAMILIES OF HETEROSOMATA.

* Pectoral fins well developed. . *Pleuronectidæ*, the *Flounders.*

** Pectoral fins wanting or rudimentary. . *Soleidæ*, the *Soles.*

FAMILIES OF PERCESOCES.

* First dorsal with 5 to 7 flexible spines; body elongated; sides with a distinct silvery band. . . ATHERINIDÆ, 99.

** First dorsal with 4 stiff spines; body compressed; no lateral band. . . . . *Mugilidæ*, the *Mullets*.

FAMILIES OF HEMIBRANCHII.

* Bones of head moderately produced; ventral fins of a stout spine and a rudimentary ray; dorsal preceded by free spines; scaleless, naked or with bony plates; an oblong, silvery, naked area in front of pectorals. . GASTEROSTEIDÆ, 98.

** Bones of head much produced, forming a long tube which bears the short jaws at the end; ventral fins without spine.

† Body compressed; no teeth; scales small; dorsal fins two; first with spines. . . *Centriscidæ*, the *Snipe Fishes*.

†† Body greatly elongated; teeth present; no scales; no dorsal spines; middle rays of caudal produced into a long filament. . . . *Fistulariidæ*, the *Trumpet Fishes*.

FAMILIES OF SYNENTOGNATHI.

* Jaws one or both elongated into a long beak.

† Both jaws elongated; no finlets; size large.
Belonidæ, the *Gar-Fishes*.

†† Lower jaw only elongate, or else dorsal and anal with detached finlets, as in the Mackerels.
Scomberesocidæ, the *Sauries*.

** Jaws moderate; pectorals elongated, nearly as long as body, used for "flying." . . *Exocœtidæ*, the *Flying Fishes*.

FAMILIES OF HAPLOMI.

* Snout depressed and elongated, its length more than half the greatest depth of body; jaws, vomer, palate and tongue armed with strong, hooked teeth; body elongated, sub-terete; size large; in fresh water. . . . . . ESOCIDÆ, 102.

** Snout rounded and rather short, its length being less than half the greatest depth of body; teeth moderate; size small.

† Vent jugular, in front of pectorals; eyes often undeveloped; ventrals small or wanting; cave fishes. AMBLYOPSIDÆ, 103.

†† Vent normal: eyes present; ventrals well developed.

‡ Margin of upper jaw formed laterally by maxillaries; lateral line inconspicuous; small dusky fishes of muddy brooks; usually a black bar at base of caudal. UMBRIDÆ, 101.

‡‡ Entire margin of upper jaw formed by intermaxillaries; head depressed; small fishes of brackish or fresh waters, often barred or striped with black.
CYPRINODONTIDÆ, 100.

FAMILIES OF ISOSPONDYLI.

* Body entirely scaleless; deep sea fishes.

† No adipose fin; throat with a long barbel; pectorals rudimentary. . . . . . *Stomiatidæ*, the *Stomiatoids*.

†† An adipose fin; no barbel; belly with phosphorescent spots.
Scopelidæ, the *Scopelids*.

** Body scaly; head naked; scales sometimes small and imbedded, sometimes large and deciduous.

‡ An adipose dorsal fin; belly rounded.

a. Margin of upper jaws formed by intermaxillaries alone.

b. Scales cycloid; deep sea fishes.
Synodontidæ, the *Synodonts*.

bb. Scales ctenoid; fresh water fishes; no teeth on vomer or palate. . . . . . . PERCOPSIDÆ, 104.

aa. Lateral margins of upper jaw formed by maxillaries; scales cycloid; in all waters. . SALMONIDÆ, 105.

‡‡ No adipose dorsal; lateral margin of upper jaw formed by maxillaries which are usually composed of three pieces; scales rather large.

c. Abdomen compressed to an edge which is serrated; lateral line obsolete; teeth very small or wanting; in all waters.
CLUPEIDÆ, 107.

cc. Abdomen compressed but not serrated; jaws, vomer and tongue with strong teeth; scales large, silvery; body compressed; lateral line well developed; fresh water.
HYODONTIDÆ, 106.

ccc. Abdomen rounded; teeth various; marine.

d. Upper jaw longest.

e. Lower jaw toothless; sides with bright silvery band.
Engraulidæ, the *Anchovies*.

ee. Both jaws with bands of villiform teeth; roof and floor of mouth with coarse patches.
Albulidæ, the *Lady Fishes.*

dd. Lower jaw longest.

f. No gular plates; no lateral line; anal short.
Dussumieridæ, the *Round Herrings.*

ff. A narrow bony plate between branches of lower jaw (much as in *Amia*). . *Elopidæ*, the *Jew Fishes.*

FAMILIES OF EVENTOGNATHI.

* Pharyngeal teeth in small number (not more than 7) and comparatively large; dorsal with 7 to 12 rays (in American species); mouth moderately or not protractile; lips scarcely or not enlarged; species mostly of small size (2 to 15 inches) Dace and Minnows. . . . . CYPRINIDÆ, 108.

** Pharyngeal teeth very numerous, small; mouth very protractile, roundish when protruded, with enlarged, fleshy lips; dorsal with 11 to 40 rays; species often of large size. Suckers.
CATOSTOMIDÆ, 109.

AA. FAMILIES OF NEMATOGNATHI.

* Margin of upper jaw formed by intermaxillaries only; maxillary rudimentary, forming the base of a long barbel.
SILURIDÆ, 110.

BB. FAMILIES OF APODES.

* Gape moderate; stomach ordinary; gill openings narrow.

† Scales rudimentary; dorsal beginning at a considerable distance behind head; fishes of shores or fresh waters.
ANGUILLIDÆ, 111.

†† Scaleless; dorsal beginning close behind base of pectorals; deep sea fishes. . . . *Congridæ*, the *Conger Eels.*

** Gape of mouth enormously wide, leading to an excessively distensible stomach; tail band-like, tapering to a fine filament. . . . . . *Saccopharyngidæ*, the *Gulpers.*

FAMILIES OF PEDICULATI.

* Head very broad; depressed.

† Skin smooth, slimy; teeth strong, card-like; dorsal VI—8, or more; the first three spines isolated, tentacle-like, on the head. . . . *Lophiidæ*, the *Fishing Frogs.*

†† Skin with conical tubercles; teeth villiform; dorsal I—4; the spine tentacle-like, retractile into a cavity beneath a prominent protuberance on forehead.

Maltheidæ, the *Sea Bats.*

** Head high, compressed; teeth card-like.

Antennariidæ, the *Diablos.*

FAMILIES OF PLECTOGNATHI.

* Jaws modified into a sort of beak, without distinct teeth.

† Both jaws divided by a median suture (teeth fused into two in each jaw); belly greatly inflatable.

Tetrodontidæ, the *Puffers.*

†† Jaws without median suture; belly scarcely or not inflatable.

‡ Body scarcely compressed, spinous, with a distinct caudal fin. . . . . *Diodontidæ*, the *Box Fishes.*

‡‡ Body much compressed, very short, truncate; the vertical fins more or less confluent.

Orthagoriscidæ, the *Globe Fishes.*

** Jaws with distinct teeth.

a. Front teeth incisor-like; 1 to 3 dorsal spines; no carapace.

Balistidæ, the *File Fishes.*

aa. Teeth slender; no dorsal spines; body enveloped in a box-like carapace, formed of hexagonal bony plates; snout, bases of fins and tail free, covered with skin.

Ostraciontidæ, the *Trunk Fishes.*

FAMILIES OF LOPHOBRANCHII.

* Tail prehensile, without caudal fin; body abruptly contracted at base of tail; head crested, out of line of axis of body.

Hippocampidæ, the *Sea Horses.*

** Tail not prehensile, with a developed caudal; body gradually tapering; direction of head in a line with axis of body.

Syngnathidæ, the *Pipe Fishes.*

ORDERS OF GANOIDEI.

* Skeleton bony; body scaly; air bladder cellular, lunglike (HYOGANOIDEI).

† Scales cycloid; snout short, broad. . CYCLOGANOIDEI, CC.

†† Scales ganoid, diamond-shaped, enamelled plates; snout lengthened, depressed. . . RHOMBOGANOIDEI, DD.

** Skeleton chiefly cartilaginous; body naked or with 3 to 5 rows of bony bucklers; vertical fins with fulcra. (CHONDROGANOIDEI.)

‡ Mouth terminal, broad; lower jaw, maxillaries and palate with many minute, deciduous teeth. SELACHOSTOMI, EE.

‡‡ Mouth narrow, inferior, toothless. . CHONDROSTEI, FF.

CC. FAMILIES OF CYCLOGANOIDEI.

* A broad bony plate between branches of lower jaw; vertical fins without fulcra; dorsal fin very long of more than 40 rays; body stout. . . . . . . AMIIDÆ, 112.

DD. FAMILIES OF RHOMBOGANOIDEI.

*Vertical fins with fulcra; dorsal short, far back, of less than 12 rays; body elongated. . . . LEPIDOSTEIDÆ, 113.

EE. FAMILIES OF SELACHOSTOMI.

* Skin naked; snout produced into a flat blade; opercle with a long flap. . . . . POLYODONTIDÆ, 114.

FF. FAMILIES OF CHONDROSTEI.

* Body with 5 rows of bony shields (rarely deciduous); snout produced; four barbels in front of mouth. ACIPENSERIDÆ, 115.

Sub-Class.—Teleostei.

(*The Bony Fishes.*)

Skeleton more or less ossified; tail homocercal; optic nerves simply crossing, without chiasma; arterial bulb simple, with two opposite valves at its origin; air bladder, if present, not lung-like; body usually scaly, sometimes covered with naked skin or bony plates; membrane bones (opercles, etc.) developed in relation to the skull. This group comprises the great majority of recent fishes.

ORDER Z.—TELEOCEPHALI.

(*The Typical Fishes.*)

This order again comprises the vast majority of recent fishes, and is characterized rather negatively, as wanting the peculiarities of the other orders than as having any positive distinctions of its own. The maxillaries are normally developed and distinct from each other, never forming the base of a long barbel. The gills are pectinated and of the ordinary pattern, and the gill-openings are in front of the pectorals and never very narrow; the subopercle is present. The scales are (when present) very rarely ossified, and are generally either ctenoid or cycloid. This group includes the *Acanthopterygians* and *Malacopterygians* of Cuvier, and the nearly corresponding *Ctenoidei* and *Cycloidei*, *Physoclysti* and

Physostomi of later writers; but however different the extremes of each (as *Percoids* and *Cyprinoids*) may be, the intervening forms are too closely related to render it possible to characterize them as distinct orders.

SUB-ORDER.—ACANTHOPTERI.

(*The Spiny-rayed Fishes.*)

FAMILY LXXXIX.—ETHEOSTOMIDÆ.

(*The Darters.*)

Fresh water fishes of small size, closely related to the *Percidæ*, but so peculiar in many respects that it seems preferable to consider them as forming a distinct family. Dorsal fins two, generally connected by membrane at the base, the second and often both dorsals high and large; anal usually well developed, with one or two spines; pectorals (except in one or two species) very large and broad, often reaching beyond base of anal; caudal large, rounded or slightly forked; scales ctenoid, sometimes absent on neck or belly, or both; head usually more or less scaly; teeth well developed on jaws and usually on vomer; eyes large; air bladder rudimentary; "sub-orbital arch incomplete." Colors often very bright; species of *Pœcilichthys* and *Diplesium* being the most brilliantly colored fresh water fishes known; sexual differences usually recognizable, the females being as a rule duller in color and more speckled or barred. In most species there is a dark streak from eye to snout, and often a dark vertical bar below the eye. Genera ten, or fewer; species about thirty, all belonging to the U. S. and Mexico, east of the Rocky Mountains, being most abundant in the Mississippi Valley, where sometimes all the genera may be found in the same stream.

Most of them prefer clear running water, where they lie on the bottom concealed under stones, darting when frightened or hungry with great velocity for a short distance by a powerful movement of the fan-shaped pectorals, then stopping as suddenly. They rarely use the caudal fin in swimming, and they are never seen moving or floating freely in the water like most fishes. When at rest they support themselves on their extended ventrals and anal. *Pleurolepis*, unlike the others, prefers a sandy bottom, where, by an almost instantaneous plunge, it buries itself in the sand and remains quiescent for hours at a time, with only its eyes and snout visible. All are carnivorous, and, in their way, voracious. All are of small size, the largest (*Percina*) reaches a length of about eight inches, while the smallest (*Microperca*), which is the smallest spiny-rayed fish known, barely attains a length of an inch and a half.

* Lateral line incomplete or wanting; body normally more or less compressed; jaws nearly even, lower jaw sometimes projecting. (PŒCILICHTHYINÆ.)

† No lateral line; dorsal spines 6 or 7; fins barred. MICROPERCA, 1.

†† Lateral line present on anterior part of body.

a. Dorsal fins slightly connected at base.

b. Spinous dorsal very low, scarcely more than half as high as soft part, commonly of 8 spines; spines in adult (male) ending in little fleshy knobs, in others pointed; color plain or with black bars and lines of dots; a black shoulder spot; fins with black lines, but no red or blue shades. . . CATONOTUS, 2.

bb. Spinous dorsal not low, nearly as high as the soft part; dorsal spines 10 to 12; colors brilliant; fins (in males at least) with bright shades of red and blue.

c. Throat and breast deep blue; dark dots along the rows of scales; a dark shoulder spot; sides often with red dots, but not notably barred. NOTHONOTUS, 3.

cc. Throat and breast orange; coloration chiefly in vertical bars rarely in lines or spots; no shoulder blotch. POECILICHTHYS, 4.

aa. Dorsal fins entirely separate; dorsal spines 8 to 10; lateral line curved upward over pectorals; small, slender species, often with red spots along the sides. BOLEICHTHYS 5.

** Lateral line present, obvious throughout its course; body normally more or less cylindrical; upper jaw more or less projecting beyond the lower.

‡ Scales obvious only along lateral line, ventral region entirely naked; body much elongated, translucent; dorsal spines 10. (PLEUROLEPINÆ.) . . . PLEUROLEPIS, 6.

‡‡ Body scaly throughout, or naked only on throat or neck. (ETHEOSTOMINÆ.)

d. Mouth scarcely inferior; lower jaw but little shorter than upper; vomer with teeth.

e. Anal spine single, sometimes obscure; mouth small, horizontal; soft dorsal usually larger than anal or than spinous dorsal; the latter of 9 (rarely 10) spines; back tessellated. . . . . BOLEOSOMA, 7.

ee Anal with two distinct spines.

f. Mouth large; soft dorsal not much larger than anal or than spinous dorsal, the latter with 10 to 15 spines; sides usually with dark bars or a chain of confluent dark blotches; no red nor blue markings, nor narrow lines of dots.

g. Belly with a series of enlarged mucronate plates along the middle line (sometimes falling off, leaving a naked strip.) . . ETHEOSTOMA, 8.

gg. Belly without enlarged plates at any time. HYPOHOMUS, 9.

ff. Mouth moderate; dorsal spines about 12; body without large bars or blotches, but each scale with a dark spot, these forming fine lengthwise lines; throat with blue; fins in ♂ with red and blue. NOTHONOTUS, 3.

dd. Mouth decidedly inferior, very small.

h. Vomer roughish, but without teeth; head very short and rounded; muzzle blunt; cheeks swollen; soft dorsal larger than anal; dorsal spines 10 to 14; sides with green blotches or markings. . DIPLESIUM, 10.

hh. Vomer with teeth; head long, pointed; the muzzle conic, truncate at tip, projecting like a hog's snout; dorsal spines about 13; sides with dark bars, alternately long and short; a small black spot at base of caudal. PERCINA, 11.

1. *MICROPERCA,* Putnam. LEAST DARTERS.

1. ***M. punctulata,*** Putnam. LEAST DARTER. Greenish olive, sides with irregular dark bars and zigzag markings; dusky lines along the rows of scales; a dark shoulder blotch; a black streak forward from eye and a vertical bar below it; D. VI to VII—9 to 12; A. II, 5 or 6; length 1¼ inches. Western and Southern States.

2. *CATONOTUS,* Agassiz. LINED DARTERS.

1. ***C. flabellatus,*** (Raf.) Putnam. FAN-TAILED DARTER. Olivaceous, dusky above; sides with obscure dusky bars; each scale with a brownish spot, these sometimes forming series of longitudinal lines but never *very* distinct ones; head narrow; mouth oblique; body rather slender; D. VIII—12; A. II, 8; length 2½ inches. Great Lakes and streams from N. Y., S. and W., abundant. (*E. linsleyi*, H. R. Storer. *Oligocephalus humeralis* and *Catonotus fasciatus*, Girard.)

2. ***C. lineolatus,*** Agassiz. STRIPED DARTER. Olivaceous, each scale with a black spot, hence the body with a series of fine dotted longitudinal lines which are very distinct above; some (♀?) further marked with dark cross bars; D. VIII—12; A. II, 8. Great Lakes and Mississippi Valley, rather northward; body deeper and colors much brighter than in the preceding, of which it is probably a variety.

3. *NOTHONOTUS,* Agassiz. TROUT DARTERS.

1. ***N. niger,*** (Raf.) Jordan. TROUT DARTER. BLUE-BREASTED DARTER. Dark olive; head blackish above; breast and throat deep blue; sides greenish, sprinkled with carmine dots, much as in a trout; series of olivaceous lines along the rows of scales; vertical fins chiefly orange at base with a bright blue edging, in ♀ merely speckled; lateral line nearly complete; D. XII—13 (D. X, Raf. Kirt.); A. II, 8; length 2½ inches. Ohio Valley, not common, one of the handsomest of our fishes; the coloration is often quite dark. [*C. maculatus*, (Kirt.) Jordan.] I have hitherto referred this genus to *Catonotus*, but the two genera must either be kept separate or both united with *Pœcilichthys*.

2. ***N. punctulatus,*** Ag. DOTTED DARTER. Greenish, mottled with black; fins all dotted; D. IX—13; A. II, 8. Mo.; Ark. (*Boleichthys whipplei*, Grd.)

4. *PŒCILICHTHYS,* Agassiz. RAINBOW DARTERS.

= *Astatichthys*, Le Vaillant.

1. ***P. cœruleus,*** (Stor.) Ag. BLUE DARTER. RAINBOW FISH. BLUE JOHNNY. Olivaceous, tessellated above, the spots running together into blotches; back without black lengthwise stripes; sides with about twelve indigo blue bars running obliquely downward, most distinct behind, separated by rich orange interspaces; caudal deep orange, edged with bright blue; anal orange, with deep blue in front and behind; soft dorsal chiefly orange, blue at base and tip; spinous dorsal crimson at base, then orange, with blue edgings; ventrals bluish, often deep indigo blue; cheeks blue; throat and breast orange, these two shades very constant; ♀ much duller, with but little or no blue or orange, the vertical fins barred or

checked; colors fade in alcohol; body short and stout; head large; D. X—12; A. II, 7; lat. l. 45; length 2 to 3 inches. Mississippi Valley, abundant; the most gaily colored of all the Darters.

2. ***P. spectabilis,*** Ag. STRIPED BLUE DARTER. Like the preceding and equally brilliant, but larger and more compressed, and more elongate; back with distinct blackish stripes along the rows of scales, pattern of coloration similar, but the colors having a clear or bleached appearance; with the other, but less abundant; often found in muddy water where *P. cœruleus* never ventures; still it is perhaps a variety.

3. ***P. zonalis,*** Cope. ZONED DARTER. Olivaceous, golden below; six dark brown quadrate spots along the back, connected by alternating spots with a brown lateral band from which eight narrow bands encircle the belly; lower fins yellow with brown spots; spinous dorsal with a crimson band; a series of crimson spots on base of soft dorsal; black spot on opercle, occiput and base of pectorals; black bars downward from eye and forward; D. XI—12; A. II, 7; lat. l. 52. Indiana to Tenn.

5. *BOLEICHTHYS,* Girard. RED-SIDED DARTERS.

> *Hololepis*, Agassiz.

1. ***B. fusiformis,*** (Grd.) Jordan. FUSIFORM DARTER. Lateral line on 12 scales, reaching middle of first dorsal; head 3¼ in length; yellowish brown; spotted; D. VIII—9; A. II, 7; lat. l. 52. Charles R., Mass.

2. ***B. erochrous,*** (Cope) Jordan. RED-SIDED DARTER. Lateral line extending to middle of first dorsal on 12 to 16 scales; head 4 in length; sides with dusky band, interrupted by red dots; D. IX—10; A. II, 7; lat. l. 44. New Jersey.

3. ***B. exilis,*** Girard. Slender Red-Sided Darter. Lateral line on 22 to 28 scales, about reaching end of first dorsal; head $3\frac{3}{4}$ in length; caudal peduncle much elongated; olivaceous brown, with zigzag markings, sides with about ten red spots, separated by brown interspaces; first dorsal bluish at base, red above; other fins reddish and barred; oblique streaks downward and forward from eye; D. IX or X—9 to 10; A. II, 7; lat. l. 60; length 2 inches. Wisconsin to Idaho; abundant in clear brooks.

6. *PLEUROLEPIS,* Agassiz. Pellucid Darters.

1. ***P. pellucidus,*** Agassiz. Sand Darter. Depth 6 to 8 in length; body nearly cylindrical; head elongated, pointed; scales small, finely dotted, far apart, and deeply imbedded; fins small; color pinkish white, pellucid in life, with a series of small, squarish, olive (blue) blotches along back, and another along sides, the spots connected by a gilt line; D. X—9; A. II, 6; length 2 to 3 inches. Ohio Valley, in sandy streams; one of the most interesting of our fishes.

7. *BOLEOSOMA,* DeKay. Tessellated Darters

> *Arlina* and *Estrella*, Grd.

* Dorsal with 9 spines; soft dorsal much larger than anal. (*Boleosoma.*)

1. ***B. olmstedi,*** (Storer) Ag. Tessellated Darter. Fins stouter and higher than in the next; depth $5\frac{1}{4}$ in length; head $4\frac{1}{4}$; olivaceous, fins barred; back tessellated; a black streak forward, and another downward from eye; cheeks and opercles scaly; neck and throat bare; D. IX—14; A. I, 8; lat. l. 50. New England to Wisconsin; abundant eastward. Western specimens generally have D. IX, 12, and perhaps vary into the next. (*B. tessellatum*, DeK.)

Var. ***atromaculata,*** (Grd.) has the neck and throat scaly.

2. ***B. effulgens,*** (Girard) Cope. "LITTLE JOHNNY." SLIM DARTER. Paler and more distinctly tessellated; brownish yellow; upper surface dotted with brown, the spots forming a few dark bars on back; a dark line from eye to snout; and sometimes a bar below eye; smaller and slenderer than the preceding, with smaller fins; eye as long as the narrow, pointed snout; cheeks, opercles, neck and throat naked; D. IX—12; A. I, 8 to I, 10; lat. l. 45. E. U. S., abundant, west of the Alleganies.

** Dorsal with 10 spines; anal as large as soft dorsal. (*Cottogaster*, Putnam.)

3. ***B. tessellatum,*** Thompson, *nec* DeK. VERMONT DARTER. L. Champlain.

8. *ETHEOSTOMA,* Rafinesque. BLACK-SIDED DARTERS.

< *Hadropterus*, Ag.

> *Diplesion* and *Alvordius*, Grd.

1. ***E. blennioides,*** Kirtland (*nec.* Raf.) BLACK-SIDED DARTER. BLENNY DARTER. Head long, pointed, 4 in length; depth 5 to $5\frac{1}{2}$; belly with a series of caducous plates along the middle line (shed at some seasons.) Straw yellow, with dark tessellations and about seven large blotches along the sides, partly confluent, thus forming a moniliform band; D. XIII, to XV—12; A. II, 9. Ohio Valley and Great Lakes; one of the most curious and elegant of all the Darters. (? *E. macrocephalum*, Cope.) (*Alvordius* and *Hadropterus maculatus*, Grd.)

2. ***E. peltatum,*** Stauffer. SHIELDED DARTER. Head shorter; sides with broad, brownish shades; ventral shields larger; D. XII—13; A. II, 9; lat. l. 53. Penn.; probably a variety of the preceding.

3. ***E. nigrofasciatum,*** (Agassiz) Jordan. Barred Blenny Darter. Salmon yellow, tessellated above; sides with nine blackish vertical bars, broad and short, almost rectangular, ending abruptly below and fading above into the dark of the back; a black spot on last rays of spinous dorsal; chest bare; second dorsal smaller than spinous dorsal or than anal, the latter extending far back; forward and downward streaks from eye; D. XI—10; A. II, 9; length 2½ inches. Indiana to Ark. and S., a pretty species, very distinct from any of the foregoing. (*Hadropterus shumardi*, Grd.)

9. *HYPOHOMUS,* Cope. Orange-Sided Darters.

1. ***H. aurantiacus,*** Cope. Orange Darter. Bright yellow, with a black lateral band formed of confluent spots in front; a few brown spots on back; fins plain; no ventral plates; D, XV—15; A, II, 11; size large. Virginia and S.

10. *DIPLESIUM,* Rafinesque. Green-Sided Darters.
= *Hyostoma*, Agassiz.

1. ***D. blennioides,*** (Raf.) Jordan. Green-Sided Darter. Olive green and tessellated above; sides with a series of about seven double transverse bars, each pair forming a **Y**-shaped figure; these are joined above, making a sort of wavy lateral band; in life, these markings are of a clear deep green; sides sprinkled with orange dots; head with olive stripes and the usual dark bars; first dorsal dark orange brown at base; blue above becoming pale at tip; second dorsal and anal of a rich blue green with some reddish; caudal greenish; young specimens much duller, but the peculiar pattern is unmistakable; body stout; head short and thick; D. XII to XIII—13; A. II, 8; length 3 inches. Penn. to

Kas. and S., abundant in Indiana, one of the handsomest of fishes. [*H. cymatogrammum*, (Abbott) Cope.] (? *E. variatum*, Kirtland.) (*H. blennioperca*, Cope.)

2. ***D. simoterum,*** (Cope) Copeland. SNUB-NOSED DARTER. Body short and thick; head very short and blunt; a series of square dark green blotches along sides and another on the back; spotted above with red; belly saffron; soft dorsal chiefly blood-red; first dorsal orange-margined; D. X—11; A. II, 7; lat. l. 52; length 3 inches. Holston R.

11. PERCINA, Haldeman. LOG PERCHES. = *Pileoma*, DeKay.

1. ***P. caprodes,*** (Raf.) Girard. LOG PERCH. HOG FISH. JACK PIKE. Salmon yellow or greenish, with about fifteen transverse dark bands from back to belly, these usually alternating with shorter and fainter ones reaching about to lateral line; a black spot at base of caudal; belly with a row of enlarged plates, shed at some seasons; D. XIII—12; A. II, 10. Great Lakes and Western Streams, abundant, E. to L. Champlain; the largest of the Darters, reaching a length of six or eight inches. (*P. zebra*, *semifasciata*, *nebulosa*, and *bimaculata* of authors.)

FAMILY XC.—PERCIDÆ.

(*The Perches.*)

Body oblong, more or less compressed, covered with rather small, strongly ctenoid scales; opercular bones mostly serrated; teeth in villiform bands on jaws, vomer and palatines; mouth slightly oblique; dorsals two, distinct, both well developed; ventrals thoracic I, 5; anal with one or two spines; branchiostegals seven; air bladder present, moderately developed; intestinal canal with a few pyloric cœca. Carnivorous fishes of moderate or rather large size, chiefly of the rivers of the Northern Hemisphere. As here restricted, a small family of about five genera and fifteen species.

* No canine teeth among the villiform ones; body compressed, cross-banded. . . . . PERCA, 1.

** With strong canine teeth; body elongated. STIZOSTEDIUM, 2.

1. *PERCA*, Linnæus. PERCHES.

1. ***P. flavescens***, (Mit.) Cuv. COMMON YELLOW PERCH. Olivaceous, sides yellowish, with broad dark bars; head $3\frac{1}{4}$ in length; depth about the same; D. XIII—14; A. II, 7; lat. l. 63. Fresh waters E. U. S., chiefly northward and eastward.

2. *STIZOSTEDIUM*, Rafinesque. PIKE PERCHES.

= *Lucioperca*, Cuvier.

1. ***S. americanum***, (Val.) Gill. WALL-EYED PIKE. YELLOW PIKE. SALMON (S.) GLASS EYE. DORY. Olive green; young yellow, with dark blotches; head $3\frac{1}{2}$ in length; depth about 4, varying with age; a dark spot on spinous dorsal behind; D. XIV—I, 20. A. II, 13; lat. l. 100. Great Lakes and Western Rivers, reaching a weight of 15 lbs. or more.

2. ***S. griseum***, (DeKay) Milner. SAUGER. GRAY PIKE PERCH. Paler; fins all spotted; a smaller spine above and one below the large opercular one; fins all black-spotted; D. XIV—I, 18; A. II, 12; lat. l. 105. Great Lakes; smaller than the preceding; perhaps the same as the next.

3. ***S. salmoneum***, Raf. WHITE SALMON OF THE OHIO. Whitish, faintly barred; much paler than *S. ameri-*

canum; dorsal fins high; head $3\frac{3}{4}$ in length; depth $5\frac{1}{2}$; D. XIV—I, 20; A. II, 11; lat. l. 100. Ohio R. "A handsome fish, resembling an *Aspro.*" (*Cove.*)

FAMILY XCI.—LABRACIDÆ.

(*The Bass.*)

Percoid fishes with the general characters of the preceding family, but having three anal spines, and the spines of the dorsal reduced in number, generally nine or ten; teeth on the tongue in our species. Genera fifteen; species thirty-five. Mostly of northern regions, the majority of them marine, often entering rivers.

* No teeth on tip of tongue; 3d anal spine longest and usually stoutest; anal with 10 to 12 soft rays. . . ROCCUS, 1.

** Teeth on tip of tongue; second anal spine stoutest; anal with 9 soft rays; opercular scales large. . . MORONE, 2.

1. ROCCUS, Mitchill. STRIPED BASS.

< *Labrax*, Cuvier.

* Body little compressed; depth less than $\frac{1}{3}$ of length; chiefly marine. (*Roccus.*)

1. ***R. lineatus,*** (Bl. & Schn.) Gill. STRIPED BASS. ROCK FISH. Silvery or yellowish, with seven or eight longitudinal bands; D. IX—1, 12; A. III, 11; lat. l. 62. Atlantic Coast, sometimes entering rivers.

** Body much compressed; depth more than $\frac{1}{3}$ of length; fresh waters. (*Lepibema*, Raf.)

2. ***R. chrysops,*** (Raf.) Gill. WHITE BASS. STRIPED LAKE BASS. Silvery, with six or more dark stripes, sometimes "so interrupted and transposed as to appear like ancient church music." D. IX—1, 12; A. III, 13; lat. l. 55. Great Lakes, Upper Mississippi Valley and N. (*Labrax multilineatus*, *notatus*, *albidus* and *osculatii* of authors.)

2. MORONE, Mitchill. WHITE BASS.

< *Labrax*, Cuvier.

1. ***M. americana,*** (Gmel.) Gill. WHITE PERCH. Whitish, usually faintly striped; depth 3 in length; D. IX—I, 12; A. III, 9; lat. l. 50. Atlantic Coast, abundant also in fresh water ponds, etc., coastwise; variable. (*L. mucronatus*, *rufus* and *pallidus*, authors.)

2. ***M. interrupta,*** Gill. SHORT-STRIPED OR BRASSY BASS. Silvery, with interrupted black stripes; D. IX—I, 13; A. III, 9. Mississippi Valley, chiefly southward. (*L. chrysops*, Grd., not of Gill.)

FAMILY XCII.—ICHTHELIDÆ.

(*The Sun Fishes.*)

Percoid fishes with a single dorsal fin, either continuous or deeply divided, with eight to twelve spines; anal fin large, with three to nine spines; ventrals thoracic, I, 5; body oblong, more or less elevated, sometimes much compressed; opercular bones feebly if at all serrated, often with entire edges; scales scarcely ctenoid, sometimes cycloid; cleft of mouth more or less oblique, lower jaw the longer; villiform teeth on jaws, vomer and usually on palatines; many species with a small supernumerary bone lying behind the maxillary and parallel with it; others with a more or less prolonged flap extending backwards from the upper angle of the opercle; nearly all with a black spot at this point, which also covers the flap if the latter is developed; colors usually brilliant, chiefly olive green, with spots or shades of blue, yellow, orange or violet. Fresh water fishes; many of them build nests which they defend with much courage; all are carnivorous, voracious and "gamey." Genera about fifteen; species forty; all American, and most abundant in the Mississippi Valley, every where forming a characteristic feature of our fish-fauna. The genera are quite well known, and most of them are firmly established; but the species of some groups, particularly *Ichthelis* and *Chœnobryttus* are in a state of almost inextricable confusion.

* Dorsal fin long, deeply divided, sometimes almost into two fins; dorsal spines 10, anal 3; opercle emarginate behind; caudal emarginate; an additional maxillary bone; mouth wide; body moderately elevated. (MICROPTERINÆ.)

MICROPTERUS, 1.

** Dorsal fin continuous, without deep division, about equal to the anal in extent; opercle emarginate behind, ending in two flat points; a slender supernumerary bone attached along the posterior margin of the broad flattish maxillary; soft fins high, mottled. (CENTRARCHINÆ.)

a. Spinous dorsal longer than soft part, forming an angle with it; dorsal spines 11 or 12, anal spines about 8.

CENTRARCHUS, 2.

aa. Spinous dorsal shorter than soft part, continuous with it; dorsal spines 6 to 8; anal spines normally 6. POMOXYS, 3.

*** Dorsal fin without division, notably larger than anal. (ICHTHELINÆ.)

† Anal with 5 or 6 spines; dorsal with 10 to 12; opercle emarginate behind; a supernumerary maxillary bone.

Scales ctenoid; caudal emarginate; fins mottled; anal spines normally 6. . . . AMBLOPLITES, 4.

Scales cycloid; caudal rounded; fins plain; anal spines normally 5. . . . ACANTHARCHUS, 5.

†† Anal with 4 spines; dorsal with 8; caudal rounded.

HEMIOPLITES, 6.

††† Anal with three spines.

b. Dorsal spines 9 (sometimes abnormally 8 or 10); caudal rounded; opercle emarginate behind—not ending in rounded flap. . . . . ENNEACANTHUS, 7.

bb. Dorsal spines 10 (rarely 9 or 11 in abnormal specimens.)

c. Opercle emarginate; caudal rounded; spinous and soft dorsals forming an angle; sides with black vertical bars. . . . . MESOGONISTIUS, 8.

cc. Opercle entire and rounded behind, often more or less produced, and always with a black spot; caudal emarginate.

d. Maxillary rather broad and flat, somewhat triangular-elongate, with (always?) a narrow supernumerary bone lying along its posterior margin; mouth wide.

e. Tongue with a conspicuous patch of teeth; mouth very large, maxillary reaching to opposite eye; opercular spot large; body deep, thick and strong, with large scales; aspect and dentition of *Amblo-plites.* . . . . . . GLOSSOPLITES, 9.

ee. Tongue without teeth; mouth smaller; maxillary not reaching to middle of eye; opercular spot small; body elongated; scales small, usually with blue dots; aspect and dentition of *Ichthelis.* CHÆNOBRYTTUS, 10.

dd. Maxillary narrow and thicker, approaching club-shaped, without supernumerary bone; mouth usually smaller; colors mostly bright.

f. Lower pharyngeal teeth conic, acute; opercular flap various, usually more or less elongated in adult. ICHTHELIS, 11.

ff. Lower pharyngeal teeth rounded, pavement-like; opercular flap rather short and broad, distinctly tipped below and behind with scarlet. POMOTIS, 12.

1. **MICROPTERUS,** Lacepede. BLACK BASS.

> *Huro and Grystes*, C. & V.
= *Calliurus*, Raf. (not of Agassiz.)
= *Lepomis*, Raf.

1. ***M. nigricans,*** (Cuv.) Gill. LARGE-MOUTHED BLACK BASS. OSWEGO BASS. Dull olive green, more or less spotted when young but not barred; usually with an

irregular dark lateral band, and three oblique stripes on opercles; end of caudal fin blackish, these markings growing obscure with age; 3d dorsal spine twice as high as first; notch between spines and soft-rays deep; eleven rows of scales between lateral line and dorsal; anal fin somewhat scaly; mouth very wide; D. X, 14; A. III, 12; lat. l. 70 to 80. Great Lakes and rivers of the West and South, abundant in most regions, and, like the next, highly valued as a food fish. (*Huro nigricans* C. & V. *G. nobilior* and *nigricans*, Agass.)

2. ***M. salmoides,*** (Lac.) Gill. Small-Mouthed Black Bass. Moss Bass. Dark green; young brighter and more or less barred and spotted, but without lateral band; tail yellow at base, then black, and edged with white; opercle with oblique olivaceous streaks; third dorsal spine half larger than first; dorsal notch rather shallow; scales larger than in the preceding—eight rows between lateral line and dorsal; mouth smaller; anal nearly scaleless; fin rays as above; lat. l. 60 to 70. Great Lakes and streams from L. Champlain S. and W.; common in N. Y. and in most regions west of the Alleganies; introduced eastward. [*G. fasciatus*, (Les.) Ag. *C. obscurus*, DeK. (young.) *M. achigan*, (Raf.) Gill, etc., etc.]

2. CENTRARCHUS, Cuvier. Many-Spined Bass.

1. ***C. irideus,*** (Bosc.) C. & V. Shining Bass. Bright green with dark spots; vertical fins mottled; dorsal with black spot behind, which is sometimes ocellated with orange; form much as in next genus; D. XII, 14; A. VIII, 15; L. 6. Rivers of Southern States.

3. POMOXYS, Rafinesque. High-Finned Bass.

* Dorsal spines normally 7; body much elevated; depth half length. (*Hyperistius*, Gill.)

1. ***P. hexacanthus,*** (C. & V.) Agass. GRASS BASS. SIX-SPINED BASS. Depth 2 in length; head nearly 3; snout projecting, forming an angle with the descending profile; mouth large, very oblique; fins very large; anal larger than dorsal; bright olive green and silvery; sides and fins much mottled; D. VII, 15; A. VI, 18; lat. l. 41; L. 8. Great Lakes to Delaware R. (*Abbott*) and S. W.; a handsome fish.

** Dorsal spines normally 6; body less elevated; depth about one-third length. (*Pomoxys.*)

2. ***P. annularis,*** Raf. BACHELOR (Ohio R.) NEW LIGHT (Ky.) CRAPPIE (St. Louis). Depth two-fifths to one-third length, scarcely greater than length of head; olivaceous, silvery below; sides with irregular clusters of dusky spots; D. VII (V to VIII), 15; A. VI, 17; lat. l. 43 (39 to 48); L. 10. Mississippi Valley; a food fish of some value; abundant and exceedingly variable. [*P. nitidus*, Grd. *P. storerius*, (Kirt.) Gill. *P. intermedius*, *protacanthus* and *brevicauda*, Gill.]

4. *AMBLOPLITES,* Rafinesque. ROCK BASS.

< *Centrarchus*, Cuvier.

1. ***A. rupestris,*** (Raf.) Gill. ROCK BASS. GOGGLE EYE. RED EYE. Depth about half length; head more than one-third; eye nearly 4 in head, very large; cheeks scaly; front convex; longest dorsal ray two-thirds depth of head at front of orbit; brassy olive with golden green and blackish markings; a dark spot at base of each scale, which is conspicuous after death, giving a striped appearance; D. XI, 11; A. VI, 10; lat. l. 42; L. 8. Great Lakes and rivers west of the Alleganies; an abundant species. [*A. æneus*, (C. & V.) Ag. *A. ichtheloides*, (Raf.) Ag.]

2. ***A. cavifrons,*** Cope. Depth $2\frac{1}{2}$ in length; mouth

larger, muzzle more projecting, the front therefore concave, cheeks nearly naked; eyes still larger, 3 in head; longest dorsal ray equal to depth of head at front of orbit; silvery, dusted with dark points; scales with dark shades; D. X, 12; A. VI, 11; lat. l. 38. Roanoke R.

5. ***ACANTHARCHUS,*** Gill. Bass Sun Fishes.

< *Centrarchus*, Baird.

1. ***A. pomotis,*** (Baird) Gill. Mud Sun Fish. Elliptical; mouth large; dark greenish olive, with dull yellowish markings and longitudinal dusky streaks; opercles with dusky radiating bars; D. XI, 12; A. V, 12; L. 5. Muddy streams coastwise, N. J., N. Y., etc.

6. ***HEMIOPLITES,*** Cope. Four-Spined Sun Fish.

1. ***H. simulans,*** Cope. Four-Spined Sun Fish. Head $2\frac{3}{4}$ in length, depth $2\frac{1}{8}$; eye $3\frac{1}{4}$ in head; bright olive, with dusky stripes; sides and cheeks with purple reflections; D. VIII, 11; A. IV, 10; lat. l. 30. James R., Va. Resembles *E. obesus*.

7. ***ENNEACANTHUS,*** Gill. Nine-Spined Sun Fishes.

< *Bryttus*, Putnam.

1. ***E. obesus,*** (Baird) Gill. Spotted Sun Fish. Depth about half length; dark olive green, much barred or mottled; cheeks with lines and spots; opercular flaps velvet black, bordered with purple; a dark bar below eye; ♂ brightly colored; fins high, spotted or mottled with blue; D. IX, 10; A. III—10; L. 3. Streams coastwise from Mass., southward; a handsome little fish. *E. guttatus* (Morris) Cope, is the male (*Abbott.*)

8. ***MESOGONISTIUS,*** Gill. Black-Banded Sun Fish.

1. ***M. chætodon,*** (Baird) Gill. Black-Banded Sun

Fish. Suborbicular; depth about half length; spines long, longest equal to distance from snout to preopercle; eye large, three in head; dirty straw color, clouded with olive; sides with four to six well-defined black vertical bars running up on the fins; first through eye, last at base of tail; D. X, 11; A. III, 12; lat. l. 28; L. 3. New Jersey to Maryland, in sluggish waters; a small, handsome species, known at once by its peculiar coloration resembling some *Chætodonts.*

9. GLOSSOPLITES, Jordan. Black Sun Fishes.

< *Calliurus*, Ag.

1. ***G. melanops,*** (Grd.) Jordan. Black Sun Fish. Depth nearly half length; head two-fifths; eye large; body very robust, broad forwards, compressed behind; nape rounded; a depression over eye, the snout projecting, and forming an angle; fins rather low, with stout spines; mouth very large, with strong teeth, the maxillary bone very broad and flat, reaching to posterior margin of eye; tongue with one or more conspicuous patches of teeth; teeth on palatines; scales large; coppery olive, or blackish, sides with faint streaks along the rows of scales; five dark bands radiating from eye across cheek; opercular spot very large; soft rays of fins barred; D. X, 10; A. III, 9; lat. l. 40. Illinois R. (Prof. *Forbes*) to Texas and W.; a fine large species having the appearance and dentition of *Amblopiites*, but with three anal spines and a rounded operculum as in *Chænobryttus.* (*Ch. charybdis*, Cope.)

2. ***G. gillii,*** (Cope.) Jordan. Red-Eyed Bream. Light green with olive cross-bars, which embrace pale spots, forming a chain-like pattern; four radiating bands behind the eye; fins blackish, barred; opercular spot small;

maxillary reaching to beyond eye; head = depth, $2\frac{3}{5}$ in length; eye large, 3 in head; D. X, 10; A. III, 9; lat. l. 45. Va. and S.; chiefly E. of the Alleganies. (Perhaps a *Chænobryttus.*)

10. CHÆNOBRYTTUS, Gill. RED EYES.

Telipomis, Raf. (Perhaps the name to be adopted.) = *Calliurus*, Ag. (but not of Raf.)

* Dorsal spines low; scales small, with blue spots.

1. ***C. cyanellus,*** (Raf.) Jordan. BLUE-SPOTTED SUN FISH. Dark clear olive-green, each scale with a blue spot and more or less of gilt edging, the body thus appearing more or less striped along the rows of scales; colors variable, golden olive, green or even almost blue; cheeks with bright blue horizontal stripes; body more or less sprinkled with irregular dark dots; vertical fins marked with green and blue and more or less edged with pale orange; usually no distinct black dorsal spot; opercular spot small, with brassy edgings; depth $2\frac{2}{3}$ in length, head $2\frac{1}{2}$; dorsal spines low and stout; pectoral fins short, $4\frac{1}{2}$ to 5 in total length (with caudal); lat. l. 48; L. 4 to 6. Ohio Valley and W. abundant. (*C. longulus* and *C. formosus*, Grd.)

Var. ***melanops,*** (Raf.) Jor. BIG-MOUTHED SUN FISH. Very near the preceding, but perhaps a distinct species; nape less elevated, body deeper, spines slenderer, mouth rather smaller; pectorals long, 3 to 4 in total length; color varying from pale olive to almost black; sides usually showing vertical bars; a black blotch on back of last rays of dorsal, and usually of anal also, as in *Ichthelis incisor.* Ohio Valley and N. W., abundant. (*C. mineopas*, Cope).

** Dorsal spines rather long; scales larger, without distinct blue spots.

2. ***C. gulosus,*** (C. & V.) Jordan. WIDE-MOUTHED SUN FISH. Olive and yellow, sides with vertical bars; no dorsal spot; form broad ovate; eye $4\frac{1}{2}$ in head. Southern States. An imperfectly described species, unknown to me.

3. ***C. nephelus,*** (Cope) Jordan. CHAIN-SIDED SUN FISH. Bright steel blue, with many bronze orange spots, which cover nearly the whole surface, so arranged that the ground color forms a series of vertical chain-like bars, very conspicuous in life; vertical fins mottled with bronze, and usually more or less edged with pale orange; no black dorsal spot; no distinct blue cheek stripes, but sides of head shaded with purplish; body rather elongate; head somewhat acute; opercular spot small; depth $2\frac{1}{3}$ in length; head nearly 3; lat. l. 42; L. 3 to 4.. Ohio Valley. A small species and one of the handsomest; extremely hardy in aquaria, and perhaps the most voracious of the Sun Fishes. (This may prove to belong to the next genus.)

* 11. *ICHTHELIS,* Rafinesque. SUN FISHES.

= *Lepomis*, Cope, and in part of Raf. (1818, not 1820).
> *Bryttus*, C. & V. (species with palatine teeth).
< *Pomotis*, Cuvier, (species without palatine teeth).

1. ***I. incisor,*** (C. & V.) Holbr. BLUE SUN FISH. COPPER-NOSED BREAM. Olive-green, adults dark, young

* The species of this genus are very closely related; the general structure (dentition, scales, fin-rays, etc.) is essentially the same in all, while those characters which usually afford specific distinctions are in the highest degree variable. The form of body, size and form of opercular flap, coloration, etc., vary very much with age, and even of specimens of the same age it is almost impossible to find two which are alike in these respects. The following descriptions are chiefly drawn up from Ohio River specimens, and they will be found to be descriptions of individuals rather than of species, and more or less deviation from them must be expected. It is hoped, however, that by their aid the student will generally be enabled to identify *adult* specimens. The study of Sun Fishes under two inches long, is, in the present state of our knowledge, extremely difficult and unsatisfactory.

more or less silvery, often uniformly so in spirits; a more or less distinct purple lustre in life; sides with undulating, sometimes chain-like, transverse bars, most conspicuous in the young; a *black spot on base of dorsal and anal behind; no blue stripes on cheeks;* no red markings; opercular flap moderately long and wide in adults, without pale edge, very short in young; body deep, compressed, caudal peduncle long and slender; head 3 in length; depth about 2; *dorsal spines very long;* D. X, 11; A. III, 10; lat. l. 40 to 45; L. 8. Great Lakes to Delaware R. (*Abbott*) and S., abundant. A large and very variable species, but almost always recognizable by the characters above emphasized. (*L. ardesiacus*, *L. megalotis*, and *L. purpurascens*, Cope.)

Var.? ***speciosus,*** (B. & G.) Jordan. SOUTHERN SUN FISH. Dorsal spines longer than the soft rays; otherwise similar to *incisor*. S. W. (*L. longispinis*, Cope. *P. heros*, Grd.)

Var.? ***obscurus,*** (Ag.) Jordan. DUSKY SUN FISH. Like *I. incisor*, but uniformly dusky; face and jaws lead color; body more elongate, and profile steeper; ear-flap long; dorsal spines rather short. Tennessee R.

2. ***I. macrochira,*** Raf. GILDED SUN FISH. *Pale* olive, young almost translucent; *sides and fins profusely speckled with golden orange, forming bars or chains;* orange below; cheeks with narrow blue horizontal lines; no dorsal spot; *flap moderate, rather narrow, narrowly bordered by light; forehead regularly convex, the bulk of the body thrown forward;* dorsal spines moderate; pectorals nearly reaching anal; ventrals elongate; head 3 in length, depth $2\frac{1}{3}$; eye = flap, 4 in head; lat. l. 42. Ohio Valley and W., abundant.

3. ***I. bombifrons,*** (Ag.) Bliss. Round-Faced Sun Fish. Light brown; fins pale, unspotted; belly and sides dotted with golden orange; head much broader, deeper and shorter than in any other species; the *profile being exceedingly prominent*, the forehead strongly arched, and the greatest depth immediately over the opercle; *flap very short and small;* soft rays of dorsal much higher than spines; depth $2\frac{1}{8}$ in length, head nearly 3. Tennessee R.; rare.

4. ***I. inscriptus,*** (Ag.) Bliss. Blue-Green Sun Fish. Dark olive green, with blue shades; *many scales marked each with a short horizontal black line*, like a pencil mark, these forming faint stripes along the sides of the back; cheeks with blue lines; opercular flap *moderate, rather narrow, directed quite obliquely upwards*, bordered above and below by pinkish; spines long; depth $2\frac{1}{2}$ in length. A small handsome species, abundant in the Valley of the Ohio and S.

5. ***I. anagallinus,*** (Cope.) Bliss. Red-Spotted Sun Fish. Dusky bluish, with greenish mottlings; *sides with many distinct, rather large, salmon-red spots;* belly bright salmon-red; opercular flap *rather large*, with a *very wide red margin, which entirely surrounds the black;* sometimes a black dorsal spot (?); spines rather high; depth $2\frac{1}{2}$ in length; scales large; lat. l. 33 to 36. Michigan to Tenn. and Kas.; a small highly colored species. (*L. peltastes*, Cope. ? *L. oculatus*, Cope.)

6. ***I. megalotis,*** Raf. Long-Eared Sun Fish. Dark olive, with blue shades; belly and sides of head strongly tinged with orange; cheeks and sides of head with blue horizontal bands; a broad blue stripe in front of eye; no dorsal spot; caudal usually dusky; *opercular flap extremely long in adults*, more or less *pale-edged*, very

variable; *dorsal spines moderate*, the longest longer than from snout to middle of eye; *body deep*, profile very steep; head, with flap, nearly half length, depth about the same; lat. l. 35 to 40. Ohio Valley, S. and W., abundant; a beautiful species. (*P. nitidus*, Kirt.)

7. ***I. fallax,*** (B. & G.) Jordan. SOUTHERN LONG-EARED SUN FISH. Very similar, but with a *distinct black spot* on last rays of dorsal at base. Lower Miss. and S. W.

8. ***I. sanguinolentus,*** (Ag.) Bliss. BLUE AND ORANGE SUN FISH. SUN PERCH. *Brilliant blue and orange*, back chiefly blue, belly entirely orange, the orange forming irregular, longitudinal rows of spots, the blue in wavy vertical lines along the edges of the scales; fins with the rays blue and the membranes orange; ventral and anal shaded with dusky blue, appearing blackish when folded; *flap very large*, with a broad pale edge which is pink behind, and bluish in front; lips blue; cheeks with bright blue and orange stripes; top of head and neck blackish; eyes bright red; *spines very low, lower than in any other species of this genus; the longest dorsal spine shorter than from snout to middle of eye:* occiput depressed; profile high and strongly curved, depth more than half length; head with flap, a little less; eye 1½ in flap; pectorals not reaching vent; lat. l. 40. Ohio Valley to Alabama. [*L. auritus*, (Raf.) Cope.] Our most brilliantly colored Sun Fish. This and the two preceding run very closely together and perhaps form one polymorphous species. If so, the name *I. auritus*, Raf., apparently intended for the present species, has priority and should be adopted unless *Labrus auritus* of Linnæus proves to have been intended for *I. rubricauda*.

9. ***I. rubricauda,*** Holbr. RED-TAILED BREAM. Dusky olive above, belly and vertical fins red; a bay spot at the base of each scale, these forming interrupted stripes; head with blue lines, especially noticeable in front of eye; *opercular flap long, rather wide, bordered with blue; spines rather high;* pectorals shorter than ventrals; depth $2\frac{1}{3}$ in length. Southern States, chiefly east of the Alleganies.

10. ***I. appendix,*** (Mitch.) Bliss. NORTHERN RED-TAILED SUN FISH. Reddish brown, with rusty red spots; cheeks with blue lines; *opercular flap long, rather narrow, scarcely pale-edged;* vertical fins more or less bright red; spines moderately high; depth $2\frac{1}{3}$ in length; pectorals as long as ventrals. Maine to North Carolina, east of the Alleganies; the only *Ichthelis* occurring in New England.

Several other species of this genus have been described within our limits, but we are unable to recognize them. A rigid reduction would perhaps show the identity of species 6, 7 and 8, and possibly *rubricauda* is a geographical variety of *appendix.*

12. *POMOTIS,* Rafinesque. POND FISHES.

1. ***P. auritus,*** (L.) Günther. COMMON SUN FISH. PUMPKIN SEED. BREAM. Depth more than half length; greenish olive above, sides orange-spotted; orange yellow below; cheeks orange with blue wavy streaks; ear-flap rounded, broadly edged with scarlet below and behind; lower fins orange, upper orange-spotted; D. X, 10; A. III, 9; lat. l. 37. Great Lakes and streams chiefly northward and E. of the Alleganies; seldom in company with *Chænobryttus.* (*P. maculatus, aureus, vulgaris,* and *luna* of authors.)

2. ***P. notatus,*** Ag. Pond Perch. More elongated; ear-flap about as above; a dusky band to eye; light olive, silvery below; spines rather slender; pectoral fins elongated; no dorsal spot; D. X, 11; A. III, 10; lat. l. 37. Headwaters of Tennessee R., abundant.

3. ***P. pallidus,*** Ag. Pale Sun Fish. Resembles *Ichthelis incisor*, but with a larger mouth; dorsal with a black spot behind; fins with dark markings; orange on ear-flap narrow; olive green, sides with eight or nine dusky bars. Tennessee R.

FAMILY XCIII.—APHREDODERIDÆ.

(*The Pirate Perches.*)

Vent jugular, in front of the ventral fins; dorsal fin single, high, with but three spines; ventrals thoracic, without spines and with more than five soft rays; some bones of head spinous; teeth on jaws and palate; scales ctenoid; branchiostegals six; cœcal appendages about twelve; air bladder simple.

A single species known from the waters of the Eastern and Southern States. It is remarkable for its voracity and for its nocturnal habits. The unusual position of the vent (as in *Amblyopsidæ*) distinguishes it widely from the Percoid fishes.

1. ***APHREDODERUS,*** LeSueur. Pirate Perches.

1. ***A. sayanus,*** (Gilliams) DeK. Pirate Perch. Head $3\frac{1}{4}$ in length; depth $3\frac{3}{4}$; greenish olive; a sub-orbital bar; caudal rounded; lower jaw longest; D. III, 11; A. II, 7; lat. l. 48; length 5 inches. N. Y. to La., in brooks near the coast.

FAMILY XCIV.—SCIÆNIDÆ.

(*The Maigres.*)

Body compressed, often elongate, covered with ctenoid scales; lateral line continuous, often running up on the caudal fin; teeth in villiform bands, sometimes with canines; vomer and palate toothless; opercles weakly if at all serrated; bones of skull more or less cavernous, with muciferous system highly developed; chin usually with pores or barbels; lower pharyngeals distinct, except in *Haploidonotus* and its relatives, where they are firmly united (pharyngognathous) as in the *Labridæ;* dorsals two, distinct or slightly connected, the soft part most developed; vertical fins usually scaly; ventrals I, 5, thoracic; anal spines generally 1 or 2; air bladder large and often complicated (rarely wanting); stomach cœcal with a few pyloric appendages.

Chiefly marine, in temperate and warm regions, the following only in fresh water. Genera about twenty; species one hundred and ten, numerous on our coasts:

* Lower jaw shorter, received within the upper; both jaws fully provided with teeth; no canines; lower pharyngeals united. (Haploidonotinæ.)

† Body moderately elevated; depth about one-third length; caudal double-truncate, slightly prolonged behind. Haploidonotus, 1.

†† Body greatly elevated at the shoulders; depth nearly half length; caudal truncate. . . Eutychelithus, 2.

1. ***HAPLOIDONOTUS,*** Rafinesque. Bubblers.

= *Amblodon*, Raf.

1\. ***H. grunniens,*** Raf. Sheepshead (Lakes), White Perch, Grunter, Drum. Depth 3 in length; head 3½; back elevated forwards, and much compressed; spines strong; first anal spine short; the second very large, attached to a stout bone; grayish silvery, dusky above; scales rather large and irregularly placed, punctate with black; D. IX — I, 30; A. II, 7; lat. l. 54. Great Lakes, Mississippi Valley, etc., abundant. (*Corvina oscula* and *C. grisea*, of Authors.)

2. ***H. concinnus,*** (Ag.) Gill. TENNESSEE DRUM. Stouter; profile steeper; dorsal beginning in advance of edge of pectorals; very large, reaching a weight of 50 lbs. Tennessee R.

3. ***H. lineatus,*** (Ag.) Gill. MISSOURI DRUM. Similar, head shorter; profile less arched; scales with darker edges, giving the body an obscurely striped appearance; very large. Osage R. This and the preceding species need confirmation.

2. *EUTYCHELITHUS,* Jordan. LAKE HURON DRUMS.

1. ***E. richardsonii,*** (C. & V.) Jordan. MALASHEGANAY LAKE DRUM. Head and shoulders much elevated; profile very steep; eye moderate; mouth rather large; the lower jaw rather projecting; head nearly one-third of length; depth about one-half; anal spine stout (single?), one-third shorter than the soft rays; pectorals pointed, much longer than the ventrals; opercular bones all finely serrated; greenish with dark bands on the back; D. IX—I, 29; A. I, 7; lat. l. 54. Lake Huron. A little known species, which, if correctly described, can not belong to *Haploidonotus*, as Prof. Gill has shown. It may be called *Eutychelithus* (Greek, *eutuches* — lucky; *lithos* — stone), from the large "ear bones," which many of the members of this family possess, and which are known to Wisconsin boys as "lucky stones."

FAMILY XCV.—COTTIDÆ.

(*The Sculpins.*)

Fishes with the cheeks mailed (*i. e.* the suborbital bone extending backward over the cheeks, articulating with the preopercle); head broad, usually not externally bony, but always more or less spinous; eyes high up, near together; body sometimes scaly, or with a series of bony plates, naked in all our species; dorsals usually two, soft dorsal largest; pectorals large, without detached rays; ventrals thoracic, near together, usually imperfect, but developed in all our species; air bladder usually absent. Genera about twenty; species about 70. Mostly of the shores of northern regions; several small species abounding in the fresh waters of Europe, Asia and North America. In habits these fresh water species bear a strong resemblance to the *Etheostomoids.* All of them are singular looking fishes, and many of the marine species are hideous in appearance.

OBS.—In the measurements given below, the length of the body is understood *inclusive* of the caudal fin, not to base of caudal as in other cases.

* Second dorsal moderately elevated, not remote from the first; preopercle with 1 to 3 spines.

† No teeth on the palate (pharyngeal teeth present, as usual); ventrals 1, 3; size small (length 2 to 3 inches). URANIDEA, 1.

†† Palate sometimes with teeth; ventrals 1, 4; size usually larger (length 3 to 6 inches). . . . PEGEDICHTHYS, 2.

** Second dorsal very high, widely separated from the first; preopercle with 4 needle-like spines. . . TRIGLOPSIS, 3.

1. *URANIDEA,* DeKay. MILLER'S THUMBS.

< *Cottus*, Girard.

* Slender, fusiform species; depth 6 to 6½ in total length.

1. ***U. gracilis,*** (Heckel) Putnam. MILLER'S THUMB. LITTLE STAR GAZER. Tips of pectorals reaching fourth

ray of second D., and first of anal; head 4 in total length; eye 4 in head; grayish, mottled. D. VIII—16; A. 11 or 12. New England and New York; the common Eastern species, found "quiescent" under stones, after the manner of the Darters. (*U. quiescens*, DeK.) [*U. boleoides* (Grd.), from Vermont, is said to be slenderer, and with larger fins. *U. formosa*, (Grd.) from stomachs of *Lota*, in deep water, L. Ontario, is more elongate, with shorter fins; it needs further examination. *U. gobioides*, (Grd.) is larger, much stouter and with larger mouth. It is from La Moille R., W. Vt.] This genus is not sufficiently distinct from the next. *Pegedichthys* is the older name.

** Stouter; depth about 5½ in total length.

2. ***U. viscosa,*** (Haldeman) Cope. Slippery Miller's Thumb. Pectorals scarcely reaching second dorsal; head 4½ in total length; eye 5 in head; body subcylindrical, covered with a viscid skin; dusky, mottled; D. VIII—17; A. 12. Streams of Penn., Md., W. Va., etc. (Youghiogheny R., *Jordan*), frequent, often found in caves. (*C. copei*, Abbott.)

3. ***U. franklini,*** (Ag.) Jordan. Franklin's Cottus. Pectorals scarcely reaching second dorsal; first dorsal scarcely lower than second; head 3⅔ in total length; eye 4⅛ in head; D. VIII—17; A. 12. S. and E. shores of L. Superior.

4. ***U. hoyi,*** Putnam, Mss. Hoy's Bull-Head. Lake Michigan. Description not yet published.

2. *PEGEDICHTHYS,* Rafinesque. Spring Fishes.

= *Potamocottus*, Gill.

1. ***P. ictalurops,*** Raf. Cave Bull-Head. Big Miller's Thumb. Cat's Eye Spring Fish. Goblin.

Head 3⅓ in length; depth 5; width of head 3½; P. reaching beyond beginning of soft dorsal to anal; preopercle with a stout erect spine and two smaller ones below; mouth wide; lateral line very distinct, chain-like, sometimes vanishing behind, and sometimes not! Grayish, mottled, three cross-blotches on back; D. VI to VIII—16; A. 12 or 13; V. 1, 4. N. C. to Ohio, Tenn., and Ind., abounding in many of the streams issuing from the caves in the limestone region; the largest species of the genus, reaching a length of 6 inches. (*C. meridionalis*, Grd. *P. carolinæ*, Gill.)

2. ***P. richardsonii,*** (Ag.) Jordan. Lake Superior Miller's Thumb. Pectorals shorter than head, scarcely reaching second dorsal; depth 6 in total length; head 4¼; eye 5½ in head; D. VIII—18; A. 14. L. Superior; one of the larger species. (*C. alvordi*, Grd., from L. Huron, "the smallest species" is said to be stouter (depth 5) and to have the pectorals longer.)

3. ***P. bairdii,*** (Girard) Jordan. Baird's Bull-Head. Pectorals long, reaching beyond beginning of anal; depth 6 to 6½ in total length; spinous dorsal very low, with a black bar; head 4¼ in length; eye 4 in head; D. VI to VII—16; A. 13. Mahoning R., Ohio; Baraboo R., Wis. (*Bundy*), etc. [*C. wilsoni*, (Grd.) from the Allegany R. is similar, but is said to have the spinous dorsal higher, and the upper rays of the pectorals branched. *U. spilota*, Cope, from Grand Rapids, Mich., is similar but stouter, and D. VIII—17.]

3. *TRIGLOPSIS,* Girard. Lake Sculpins.

= *Ptyonotus*, Günther.

1. ***T. thompsoni,*** Grd. Deep Water Lake Sculpin. Body elongated; depth 7 in total length; head 3½; eyes very large, 4 in head; D. VI—18; A. 15. Great Lakes

(L. Ontario, L. Michigan) in deep water; till lately known only from remains found in the stomachs of Lake Trout and Ling.

FAMILY XCVI.—GOBIIDÆ.

(*The Gobies.*)

Body elongated, low, naked or scaly; dorsals two, sometimes united, the spines flexible and less developed than the soft rays; anal similar to soft dorsal; ventrals 1, 5 (rarely 1, 4), sometimes united, forming a disk; gill openings narrow; teeth generally small; a prominent papilla near the vent, as in the Blennies; air bladder usually wanting; no pyloric appendages. Genera thirty; species three hundred and twenty-five. Of the seas of temperate and tropical regions, found on the bottoms near the shore. A few species inhabit both salt and fresh water.

* Body naked; ventrals united. . . . GOBIOSOMA, 1.

1. GOBIOSOMA, Girard. NAKED GOBIES.

1. ***G. molestum,*** Grd. Dusky brown; head 3½ in length; D. VII—12; A. 12. Coast of Texas, entering rivers. A single specimen in the Museum of Comp. Zoölogy from the Ohio R., near Louisville (*Putnam*).

SUB-ORDER.—ANACANTHINI.

(*The Jugular Fishes.*)

FAMILY XCVII.—GADIDÆ.

(*The Cod Fishes.*)

Body elongated, covered with small smooth scales; dorsal fins 1, 2 or 3, occupying most of the back; rays of posterior part well developed; vertical fins never entirely united (as in some related families); ventrals jugular, usually several-rayed; gill openings wide; air bladder usually present; no pseudobranchiæ; pyloric cœca usually in large number (30 or more in *Lota*). Genera about twenty-five; species about seventy. An important family found chiefly in the northern seas; a single genus inhabiting the lakes and larger streams of the northern parts of Europe and America.

* Chin with a barbel; dorsals 2; anal single; teeth villiform. LOTA, 1.

1. LOTA, Cuvier. LINGS.

1. ***L. lacustris,*** (Mitch.) Gill. LING. BURBOT. LAKE LAWYER. EEL-POUT. LAKE CUSK. Dark olive, thickly marbled with blackish, yellowish or dusky beneath; head broad, depressed; body sub-cylindrical in front, compressed behind; upper jaw (always?) longest; D. 13—76; A. 68; V. 7; length 1½ to 2½ feet. Great Lakes and streams of New England, north to the Arctic Circle, abundant; very rare in the Miss. Valley. A curious fish, rarely used for food, although the livers are said to be delicious. [*L. maculosa*, (Les.) Cuv. *L. compressa*, (Les.) and *L. brosmiana*, Storer. *L. inornata*, DeK., etc.] It is closely related to the European *L. vulgaris*, Cuv.

SUB-ORDER.—HEMIBRANCHII.

(*The Half-Gilled Fishes.*)

FAMILY XCVIII.—GASTEROSTEIDÆ.

(*The Sticklebacks.*)

Small fishes with the body elongated and compressed; caudal peduncle very slender; mouth large with the cleft oblique; villiform teeth on jaws and pharyngeals; branchiostegals three; opercles unarmed; sub-orbital bone articulated with the preopercle (as in *Cottidæ*, with which these fishes were formerly associated); skin naked or with bony plates; dorsal preceded by two or more isolated spines; ventrals abdominal, of a stout spine, accompanied by a rudimentary ray; air bladder simple; a few pyloric cœca. Genera about four; species twenty-five or less, in fresh waters and arms of the sea in northern Europe and America.

* Dorsal with 2 to 6 free spines.

† Sides mailed; a serrated bony ventral cuirass and usually a bony caudal keel. . . . . Gasterosteus, 1.

†† Sides naked; no caudal keel; ventral cuirass reduced, not serrated.

‡ Dorsal spines not in a right line when erected, the anterior ones highest; ventral plates 2—not on median line; caudal peduncle very slender. . . Apeltes, 2.

‡‡ Dorsal spines in the same line, the lowest in front; ventral plate single, on the middle line of abdomen; caudal peduncle stouter. . . . . . Eucalia, 3.

** Dorsal spines 7 or more; sides mailed or not. Pygosteus, 4.

1. GASTEROSTEUS, Linnæus. Mailed Sticklebacks.

1. ***G. noveboracensis,*** C. & V. New York Stickleback. Maine to Cape Hatteras, coastwise; sometimes ascending streams.

2. APELTES, DeKay. Naked Sticklebacks.

< *Gasterosteus*, L.

1. ***A. quadracus,*** (Mitch.) Brevoort. Four-Spined Stickleback. Abundant, with the preceding.

3. EUCALIA, Jordan. Nest-Building Sticklebacks.

< *Apeltes*, Authors.

1. ***E. inconstans,*** (Kirtland) Jordan. Brook Stickleback. Ohio Stickleback. Head about 3½ in length;

depth nearly 4; spines rather low; ventral spine about equal to eye; color olivaceous, marbled with darker; males in spring jet black, finely punctate; D. III to V—I, 10; A. I, 10; length 2½ inches. Ohio to Minnesota and Kansas, chiefly northward; abundant in sluggish streams; an interesting species, remarkable for its pugnacity and for its nest-building habits.

Var. ***pygmæa,*** (Agassiz) Jordan. L. SUPERIOR STICKLEBACK. Depth 3¾ in length; head 3½; caudal peduncle short and stout; body shorter and deeper than the preceding; vent much nearer tip of caudal than snout; color similar; D. III or IV—I, 6; A. I, 6. L. Superior.

Var. ***cayuga,*** Jordan. CAYUGA LAKE STICKLEBACK. Head 3⅓ in length; depth 4¼; spines all high; caudal peduncle slender; vent much nearer snout than tip of caudal; D. IV—I, 10; A. 1, 10. Cayuga L., N. Y., dredged in deep water (*Wilder*). Probably this and the preceding are varieties of the variable *E. inconstans.* Length 1¼ inches.

4. ***PYGOSTEUS,*** Brevoort. MANY-SPINED STICKLEBACK.

1. ***P. occidentalis,*** (C. & V.) Brevoort. TEN-SPINED STICKLEBACK. Coastwise, abundant; sometimes ascending streams.

2. ***P. nebulosus,*** (Ag.) Jordan. MANY-SPINED LAKE STICKLEBACK. Head 3½ in length; ventral spine long; caudal keeled; sides not mailed; D. IX — 10; A. I, 8. Great Lakes (scarcely distinct from preceding.)

3. ***P. mainensis,*** (Storer) Brev. MAINE STICKLEBACK. Sides with a serrated plate; body banded; D. VII — I, 9; A. I, 8. Kennebec R., Maine.

SUB-ORDER.—PERCESOCES.

(*The Silversides.*)

FAMILY XCIX.—ATHERINIDÆ.

(*The Silversides.*)

Body elongated, more or less compressed, covered with rather small cycloid scales; sides with a bright, distinct silvery band in all known species; dorsal spines flexible and feeble; teeth small, numerous. Small, carnivorous fishes of warm regions, usually swimming in schools near the shore; a few species in permanently fresh water. Genera about five; species forty-five. Besides the following strictly inland species, the common Dotted Silverside [*Chirostoma notatum* (Mitch.) Gill] ascends rivers from the sea.

* Mouth very oblique; the upper jaw plane above, concave within; the lower jaw correspondingly convex, the protractile intermaxillaries forming a peculiar roof-like beak.
Labidesthes, 1.

1. ***LABIDESTHES,*** Cope. River Silversides.

1. ***L. sicculus,*** Cope. Silver Skip-Jack. River Silverside. Depth 6 in length; head $4\frac{1}{2}$; eye $3\frac{1}{2}$ in head; anal long, nearly one-third of length of body; scales small; pale olive, translucent, dotted with black, the silver lateral band very distinct; D. IV — 11; A. I, 23; lat. l. 75; length 3 to 4 inches. Western streams and ponds, Mich. to Ills. and Tenn.; abundant where found, but not noticed till comparatively lately. A very slender and elegant species of delicate organism. The peculiar "duck-like muzzle" is said to resemble that of some *Cyprinodonts*, especially the Central American *Belonesox*.

SUB-ORDER.—HAPLOMI.

(The Toothed Minnows.)

FAMILY C.—CYPRINODONTIDÆ.

(The Cyprinodonts.)

Head and body scaly; no barbels; margin of upper jaw formed by intermaxillaries only; teeth in both jaws and on pharyngeals well developed; dorsal fin far back; caudal usually rounded; no adipose fin; lateral line rudimentary; air bladder simple; no pyloric cœca; head more or less flattened above, the lower jaw usually longer; sexes commonly unlike, the female larger; anal of male often modified into a sword-shaped intromittent organ; chiefly viviparous.

Small fishes of fresh or brackish waters in both continents; most abundant in warm regions. Genera twenty-five; species one hundred and twenty. A recently discovered *Cyprinodont* (*Protistius*, Cope.) from S. A. is said to have a rudimentary spinous dorsal fin, indicating a close relationship between this family and the *Percesoces*.

Our numerous species are not well known, and the current genera are but indifferently characterized. One species (*Girardinus formosus*) from S. C. and Florida is said to be the smallest known vertebrate. The species here mentioned are carnivorous surface swimmers; many southern species feed on mud and slime.

* Dorsal fin commencing distinctly before anal; branchiostegals about 5. . . . . Fundulus, 1.

** Dorsal fin short, commencing behind or opposite the elongate anal; branchiostegals about 3. . . Zygonectes, 2.

1. ***FUNDULUS,*** Lacepede. Killifishes.

1. ***F. diaphanus,*** (Les.) Ag. Barred Killifish. Spring Mummichog. Sides silvery olive, with twelve

to fifteen distinct, narrow, blackish, vertical bars; head rather narrow; D. 13; A. 12; lat. l. 42. Coastwise, abundant, but ascending streams to their sources, hence found in clear springs as far inland as Mich. (*Cope*), Wisconsin (*Copeland*), Illinois, Colorado (*Yarrow*), etc. [*F. multifasciatus*, (Les.) Val.] Various marine species of this genus, as well as of *Cyprinodon*, *Hydrargyra* and *Micristius* are sometimes found in fresh waters near the coast.

2. *ZYGONECTES*, Agassiz. Top Minnows.

< *Haplochilus*, Günther.

× *Hydrargyra*, *Pœcilia* and *Fundulus* of Authors.

1. ***Z. olivaceus***, (Stor.) Ag. Black-Sided Killifish. Top Minnow. Depth $4\frac{1}{2}$ in length; head 4; head broad, depressed; clear pale olive with a few dots above; a wide purplish-black band along sides from snout through eye to caudal, its margin usually serrated; D. 9; A. 11; lat. l. 34; length $2\frac{1}{2}$ inches. Miss. Valley; abundant. (*Z. pulchellus* and *tenellus*, Grd. *F. aureus*, Cope, etc.)

2. ***Z. nottii***, Agassiz. Striped Top Minnow. A broad band and several dotted lines along sides; the darker continuous bands alternating with fainter interrupted ones; males transversely banded; silvery below. Mississippi Valley and Southern streams.

3. ***Z. melanops***, (Cope) Jordan. Yellowish brown; belly golden; a black spot below eye; fins dotted; D. 6; A. 8; lat. l. 31. Neuse R.

4. ***Z. catenatus***, (Storer) Jordan. Stud Fish. May Fish. Pale steel blue, sides with series of bright bronze spots, forming very distinct longitudinal streaks; head with bright green stripes; throat and bars on dorsal and

anal bright orange; D. 14; A. 15; lat. l. 44; length 6 inches. Tenn. R.; one of the handsomest of the family. It probably does not belong to this genus, but it is equally unlike *Fundulus*, *Hydrargyra* and *Pœcilia*, to all of which genera it has been referred.

FAMILY CI.—UMBRIDÆ.

(*The Mud Minnows.*)

Small fishes like the *Cyprinodonts* in most respects, but with the mouth different; margin of upper jaw formed by the intermaxillaries mesially and by the maxillaries laterally; head and body scaly; no lateral line; scales moderate, cycloid; lower jaw longest; dorsal far back; caudal fin rounded; gill openings wide; teeth villiform on jaws, vomer and palatines. Genus one (or two); (*Melanura* has never been properly distinguished from *Umbra*); species two, *Umbra crameri* of Austria and the following. Both are found in sluggish brooks in mud or among weeds. "A locality which, with the water perfectly clear, will appear destitute of fish, will perhaps yield a number of mud fish on stirring up the mud at the bottom and drawing a seine through it. Ditches in the prairies of Wisconsin, or mere bog-holes, apparently affording lodgment to nothing beyond tadpoles, may thus be found filled with *Melanuras*." (*Baird.*)

1. ***MELANURA,*** Agassiz. MUD MINNOWS.

< *Umbra*, Günther.

1. ***M. limi,*** (Kirtland) Agassiz. MUD MINNOW. MUD DACE. DOG FISH. Depth about 4 in length; head 3½; head rather large, flattish above; greenish or dark olive; sides with narrow pale bars, often obscure; a distinct black bar at base of caudal; D. 14; A. 9; V. 6; lat. l.

35; length 2 to 4 inches. New Jersey to Minnesota, chiefly northward and westward; most abundant in Wisconsin; rare in Ohio Valley; usually associated with *Eucalia inconstans.*

FAMILY CII.—ESOCIDÆ.

(*The Pikes.*)

Body elongated, sub-cylindrical, with rather small scales; margin of upper jaw formed by intermaxillaries mesially and by the maxillaries laterally; mouth very large; jaws elongate, depressed; teeth strong, hooked, unequal, on intermaxillaries, vomer and palatines; dorsal short, opposite anal; gill openings wide; air bladder present. Voracious fishes of the fresh waters of northern regions, two or three of the species reaching a large size. With a single exception (*E. lucius*, L. the Pike of Europe and Asia) all the species belong to the U. S., and our *E. estor* is perhaps identical with *E. lucius.*

Genus one (or two); species five to twenty; the following seem to be well characterized; many others have been described and some of them may be good, but that has yet to be proven. It may be convenient to recognize the sub-genus *Picorellus*, proposed long ago by Rafinesque.

* Lower half of opercles scaleless; cheeks scaly or not; species of large size; grayish blue in color, with round whitish spots. . . . . Esox, 1.

** Cheeks and opercles entirely scaly; size smaller; color olive green, with darker bars or reticulations; a black vertical bar below the eye. . . . . Picorellus, 2.

1. ESOX, Linnæus. Pikes.

1. ***E. nobilior,*** Thompson. Muskallunge. Great Pike. Cheeks as well as opercles half bare; grayish

with round white spots; a magnificent fish, reaching a length of 6 feet; B. 19; D. 19 to 21; A. 20; lat. l. 155. Great Lakes, etc. (*E. estor* of some authors.)

2. ***E. lucius,*** L., var. ***estor,*** (LeSueur.) Great Lake Pike. Northern Pickerel. Cheeks entirely scaly; depth 7 in length; head 3½; olive gray; sides with round yellowish spots as large as peas; each scale with a shining V-shaped mark opening downwards; B. 15; D. 20; A. 17; lat. l. 122. Great Lakes and headwaters of the Mississippi. A fine species reaching a length of 3 to 4 feet. (*E. lucius*, *lucioides*, *boreus*, etc., of authors.)

2. *PICORELLUS,* Rafinesque. Pickerels.

* Branchiostegals 14 to 16; snout prolonged; front of eye nearly midway in head.

1. ***P. reticulatus,*** (LeSueur) Jordan. Common Eastern Pickerel. Green Pike. Head 3⅛ in length; the snout much prolonged; front of eye about midway in head; eye more than three times in snout; green, sides with a network of brown streaks; B. 14 to 16; D. 16 to 18; A. 15 to 17; lat. l. 120 to 130. Streams of Atlantic States abundant, but not found far in the interior; smaller than the preceding, but much larger than the next. Represented S. of Va. by *P. affinis*. (Holbr.)

** Branchiostegals normally 12; front of eye nearer tip of snout.

2. ***P. americanus,*** (Lac.) Jordan. Banded Pickerel. Trout Pickerel. Head 3⅗ in length, the snout much shorter than in the preceding; eye much nearer snout than opercular margin, its diameter less than 3 in snout; dark green; sides with about twenty distinct blackish curved bars, scarcely reticulated; B. 12; D. 13; A. 13; lat. l. 100; length scarcely a foot. Atlantic streams, with

the preceding. (*E. niger*, *scomberius*, *fasciatus* and *ornatus* of authors.) Represented S. of Va. by *P. ravenelii*. (Holbr.)

3. ***P. salmoneus,*** (Raf.) Jordan. LITTLE PICKEREL. WESTERN TROUT PICKEREL. Size and general form of preceding or slenderer; olivaceous green above; white below; sides with many reticulations and curved streaks, instead of bars; a black streak in front of eye as well as below; B. 12; D. 13; A. 14; lat. l. 112. Western streams, abundant in the Ohio Valley. (*E. cypho*, *E. porosus*, Cope, etc.) (*E. umbrosus*, Kirtland.) Resembles *reticulatus* more than *americanus*.

FAMILY CIII.—AMBLYOPSIDÆ.

(*The Cave Fishes.*)

Fishes with the ventral fins rudimentary or wanting; the vent jugular, in front of the pectorals, and the eyes sometimes rudimentary and concealed under the skin; margins of upper jaw formed by intermaxillaries alone; head naked; body with small, cycloid scales, irregularly arranged; no lateral line; villiform teeth on jaws and palate; dorsal far back, opposite anal; stomach cœcal, with pyloric appendages; some (and probably all) viviparous.

Fishes of small size living in subterranean streams and ditches of the central and southern U. S. Three genera and four species are "all of the family yet known, but that others will be discovered and the range of the present known species extended is very probable. The ditches and small streams of the lowlands of our Southern Coast will undoubtedly be found to be the home of numerous individuals, and perhaps of new species and genera, while the subterranean streams of the central portion of our country most likely contain other species." (*Putnam.*)

* Eyes rudimentary, concealed under the skin; body colorless.

Ventrals present, small. . . . AMBLYOPSIS, 1.

Ventrals entirely wanting. . . TYPHLICHTHYS, 2.

** Eyes well developed; body colored; no ventrals.

CHOLOGASTER, 3.

1. AMBLYOPSIS, DeKay. LARGER BLIND FISH.

1. ***A. spelæus,*** DeKay. BLIND FISH OF THE MAMMOTH CAVE. Head 3 in length; D. and A. equal, well developed; head and body with papillary ridges; scales small; colorless; D. 10; A. 9; V. 4; P. 11; length 2 to 5 inches. Subterranean streams of Ky. and Ind. Mammoth Cave, Wyandot Cave, etc.

2. TYPHLICHTHYS, Girard. SMALL BLIND FISH.

1. ***T. subterraneus,*** Grd. General character of Amblyopsis, but the head rather blunter and broader forwards; D. 7 or 8; A. 7 or 8; P. 12; length 2 inches or less. Subterranean streams in Ky., Tenn., Ala.

3. CHOLOGASTER, Agassiz. DITCH FISHES.

1. ***C. cornutus,*** Ag. Head 3 in length; eye moderate, well developed; snout with two horn-like projections; yellowish brown, dark above; sides with three dark lines, becoming dots on the tail; middle rays of C. dark, fins otherwise uncolored; D. 8 or 9; A. 8 or 9; P. 12; length 2 to 2½ inches. Ditches in a rice field, Waccamaw, S. C. Three specimens known.

2. ***C. agassizii,*** Putnam. Head 4 in length; eyes larger; uniform light brown, otherwise as above; length 1 to 2 inches. Subterranean streams in Tenn. and Ky.

SUB-ORDER.—ISOSPONDYLI.

(*The Trout-like Fishes.*)

FAMILY CIV.—PERCOPSIDÆ.

(*The Trout Perches.*)

Body covered with moderate-sized ctenoid scales; head naked; no barbels; opercles well developed; gill openings wide; an adipose fin; jaws with villiform teeth; no teeth on vomer or palate; margin of upper jaw formed by intermaxillaries alone; branchiostegals six. A single genus and one or two species inhabiting the fresh waters of the northern U. S. Interesting little fishes, with the general characters of *Salmonidæ*, but having the mouth and scales decidedly Perch-like.

1. *PERCOPSIS,* Agassiz. TROUT PERCHES.

1. ***P. guttatus,*** Ag. Depth $4\frac{1}{2}$ in length; head $3\frac{2}{3}$; silvery, almost pellucid; upper parts with rounded dark spots made up of minute dots; D. 11; A. 7; L. 10. Great Lakes; Ohio R. (*Jordan*); Potomac R. (*Baird*); Delaware R. (*Abbott.*)

FAMILY CV.—SALMONIDÆ.

(*The Trout.*)

Head naked, body scaly, no barbels; margin of upper jaw formed by intermaxillaries mesially and by maxillaries laterally; adipose fin present; belly rounded; air bladder large, simple; pseudobranchiæ present; pyloric appendages usually numerous; eggs falling into the cavity of the abdomen before exclusion. Fresh waters of northern regions, many species periodically descending to the sea; a few permanently marine.

Genera sixteen; species one hundred and sixty. The

variations due to age, sex and food are very great, and have led to the establishment of a great number of nominal species in all the leading genera, particularly in *Salmo*.

* Jaws with evident teeth.

† Dorsal moderate of less than 20 rays; teeth strong, on jaws, vomer and tongue.

‡ Scales small; partly imbedded in the skin; lat. l. 100 or more. . . . . SALMO, 1.

‡‡ Scales moderate, deciduous, not imbedded; lat. l. 60 to 70. OSMERUS, 2.

†† Dorsal very high of 20 or more rays; teeth small. THYMALLUS, 3.

** Teeth wanting or reduced to slight roughnesses; scales rather large, loose.

a. Lower jaw longer than upper, or if not, body slender, subfusiform. . . . . ARGYROSOMUS, 4.

aa. Upper jaw notably longest; body more or less elevated. COREGONUS, 5.

1. SALMO, Linnæus. SALMONS.

> *Salmo*, *Fario*, *Salar*, *Trutta*, *Umbla*, *Hucho*, *Salvelini*, etc., of Authors.

* Anadromous species, running up from the sea into fresh water to spawn; the young remaining there for a time, then returning to the sea where they remain except during the season of reproduction; upper jaw in males moderately if at all hooked. (*Salmo.*)

1. ***S. salar***, L. GREAT SEA SALMON. No red spots; young (*Parr. Smolt*) with dusky cross bars; males in the spawning season with the lower jaw strongly recurved and hooked; body covered with black and red patches; others silvery, with small black dots; eleven or twelve scales in a transverse series from behind the adi-

pose fin obliquely forward to the lateral line; D. 14; A. 11; lat. l. 120. Northern Europe and America, S. to Cape Cod.

** Species not anadromous, living entirely in fresh water or only occasionally passing down to the sea. (Trout.)

† In flowing fresh water, retiring to deeper places in winter; red-spotted. (*Hucho*—part.)

2. ***S. fontinalis,*** Mitchill. Brook Trout. Speckled Trout. Mouth wide; teeth moderate; body olivaceous, variegated with blackish, with numerous red spots; lower fins usually orange with black and white marginal bands; dorsal with black spots; colors variable; young barred; D. 12; A. 12; lat. l. 200. A well known and beautiful fish, in clear brooks from the French Broad R. to the Arctic regions.

†† In deep rivers or lakes, ascending shallow streams to spawn.
a. Red-spotted.

3. ***S. oquassa,*** Grd. Blue-Back Trout. Oquassa. Slender, "the most graceful of all the trouts;" blue or bluish above; sides and below silvery in female, orange in male; sides spotted with orange in both sexes; upper fins bluish, bordered with orange; lower fins fiery orange, margined with white. Oquassa L. and other lakes in Maine.

aa. Black-spotted.

4. ***S. sebago,*** Grd. Sebago Lake Trout. Union River Trout. Everywhere black-spotted; scales quite large; D. 14; A. 10; V. 10; lat. l. 115. Sebago L., Union R., and other waters in Maine. (*S. gloveri*, Grd.)

††† Trout living in deep fresh water lakes, coming to the shores to spawn in shallow water; never entering running brooks or passing to the sea. (*Trutta.*)

5. ***S. namaycush,*** Pennant. Mackinaw Trout. Great Lake Trout. Stout; head very large, $3\frac{1}{2}$ in length;

bones of head strong; posterior point of juncture of opercle and sub-opercle much nearer the upper end of the gill opening than to the lower anterior angle of the sub-opercle; teeth strong; fins large, the caudal deeply forked; color grayish, more or less spotted, varying much with circumstances; D. 13 to 14; A. 12; V. 9; lat. l. 220 length 2 to 6 feet. All the Great Lakes, north to the Arctic Sea; a fish of much firmer flesh than the next. (*S. amethystus*, Mit.)

6. ***S. siscowet,*** Agassiz. SISCOWET. L. SUPERIOR TROUT. Stout; head smaller, $4\frac{1}{8}$ in length; posterior point of junction of opercle and sub-opercle nearer to the lower anterior angle of sub-opercle than to the upper end of gill opening; fins and teeth well developed but weaker than in *S. namaycush;* ventrals farther back; caudal less forked; flesh fat and not firm; grayish, with round white spots and markings; D. 12 to 14; A. 12; lat. l. 200. L. Superior, L. Huron.

7. ***S. confinis,*** Mit. LAKE TROUT OF NEW YORK. Blackish, with gray spots; body unusually short and thick. Lakes of Central and Western N. Y.; a doubtful species.

8. ***S. symmetrica,*** Prescott. WINNIPISEOGEE TROUT. Grayish and brown above, marbled with darker; white below; body unusually slender and symmetrical. Lake Winnipiseogee; also a doubtful species.

2. ***OSMERUS,*** Linnæus. SMELTS.

1. ***O. mordax,*** (Mit.) Gill. COMMON SMELT. Head 4 in length; eye 4 to $4\frac{1}{4}$ in head; teeth stout, especially large on the tongue; transparent greenish, a silvery band along sides; scales very loose; D. 11; A. 15; lat. l. 66. Coast, Nova Scotia to Virginia; also "land-locked" in fresh water ponds in Maine, etc. (*O. viridescens*, Mit.)

Var. ***spectrum,*** (Cope) Jordan. LAND-LOCKED SMELT. Head $4\frac{1}{2}$ in length; eye large, 3 in head; depth $8\frac{1}{3}$ in length. Wilton Pond, Maine.

Var. ***abbottii,*** (Cope) Jordan. ABBOTT'S SMELT. Head $4\frac{3}{4}$ in length; eye $4\frac{1}{2}$ in head; depth 7 in length; colors dark; lat. l. 68. Cobessicontic L., Maine.

3. *THYMALLUS,* Cuvier. GRAYLINGS.

1. ***T. tricolor,*** Cope. MICHIGAN GRAYLING. Depth $4\frac{2}{3}$ in length; head about the same; purplish gray, silvery below; dorsal with rosy markings and rows of green or blue spots; D. 27; A. 13; lat. l. 97. Waters of the north part of the S. peninsula of Michigan; a beautiful fish.

4. *ARGYROSOMUS,* Agassiz. SISCOES.

* Body sub-fusiform; depth 4 to 5 in length.

1. ***A. clupeiformis,*** (Mitch.) Ag. LAKE HERRING. MICHIGAN HERRING. Head $4\frac{3}{4}$ in length ($4\frac{1}{4}$ to $5\frac{1}{4}$); depth 4 ($3\frac{2}{3}$ to $4\frac{1}{3}$); eye 4 in head; maxillary $3\frac{1}{2}$ to $3\frac{3}{4}$; mandible $2\frac{1}{3}$; scales rather large and loose; bluish above, silvery on sides and below; D. 12; A. 13; lat. l. 76; length 12 to 18 inches. Great Lakes, etc., very abundant; a shallow water species. (*Coregonus albus, artedi, lucidus, harengus*, etc. of authors.)

2. ***A. sisco,*** Jordan. SISCO OF LAKE TIPPECANOE. Head $4\frac{1}{2}$; depth $4\frac{1}{3}$; eye $3\frac{3}{5}$ in head; maxillary $3\frac{1}{3}$; mandible $2\frac{1}{3}$; longest dorsal ray three times length of shortest; steel blue above, sides silvery but without the clear lustre of *A. hoyi*, finely punctate; D. 11; A. 13; lat. l. 84. Lakes of Indiana and Wisconsin, living in deep water except at the spawning season; very close to the preceding, of which it is probably a variety, but the habits are more like those of the next.

3. ***A. nigripinnis,*** Gill. BLACK FIN. Head $4\frac{1}{2}$ in length; depth the same; eye 4 in head; body compressed; fins blackish, darker than in the others; D. 12; A. 12; lat. l. 80; length 16 to 18 inches; a much larger fish than the preceding or the next. Lake Michigan, in deep water.

4. ***A. hoyi,*** Gill. SISCO OF LAKE MICHIGAN. Head 4 in length; depth $4\frac{1}{2}$; eye large, $3\frac{3}{4}$ in head; maxillary $2\frac{3}{4}$; mandible 2; longest ray of dorsal four times the length of the shortest; upper jaw somewhat projecting, the mouth appearing much as in *Coregonus;* bluish above, sides lustrous silvery, more brilliant than in any other species; D. 11; A. 12; lat. l. 74; length 8 inches. Smallest and handsomest of the Siscoes, in the deep waters of the Upper Lakes.

** Body elevated; depth about 3 in length.

5. ***A. tullibee,*** (Rich.) Ag. TULLIBEE. Head $4\frac{1}{4}$ in length; D. 15; A. 15; lat. l. 77. L. Superior and N.

5. *COREGONUS,* Linnæus. WHITE FISHES.

1. ***C. albus,*** LeSueur. LAKE WHITE FISH. Depth $3\frac{1}{2}$ in length; head small, $5\frac{1}{4}$; eye 4 in head, about as long as snout; form varying much with age, sex and food; pale olive above; sides white; D. 13; A. 13; lat. l. 75 to 86. Great Lakes and bodies of water tributary to them, north to the Arctic Sea. (Various other species have been described within our limits, but it is impossible to distinguish them.)

FAMILY CVI.—HYODONTIDÆ.

(*The Moon Eyes.*)

Body much compressed, covered with large, silvery cycloid scales; head naked; margin of upper jaw formed by intermaxillaries mesially and by maxillaries laterally;

no barbels; no adipose fin; lateral line distinct; abdomen not serrated, compressed; moderate sized teeth on jaws, vomer, sphenoid, hyoid, pterygoid and palatine bones; tongue with large teeth; head short, deep; eye very large; gill openings wide; one pyloric appendage; air bladder simple. A single species, inhabiting our Western Streams and the Great Lakes.

1. HYODON, LeSueur. MOON EYES.

1. ***H. tergisus,*** LeSueur. MOON EYE. SILVER BASS. TOOTHED HERRING. Depth $3\frac{1}{3}$ in length; head $4\frac{2}{3}$; snout rounded, shorter than the large eye, which is $3\frac{1}{3}$ in head; scales largest on the flanks; pale olivaceous above, sides brilliantly silvery; D. 3, 12; A. 30; V. 7; lat. l. 59; length 1 foot. Great Lakes and Mississippi Valley abundant; one of our most beautiful fresh water fishes; variable; it has been described under many names.

FAMILY CVII.—CLUPEIDÆ.

(*The Herrings.*)

Body scaly; head naked; abdomen compressed to an edge and sharply serrated; margin of upper jaw formed by intermaxillaries mesially and maxillaries laterally; maxillaries composed of three pieces which are sometimes movable; teeth usually minute or wanting; dorsal moderate; anal often very long; scales usually large and loose; no lateral line; gills well developed; posterior part of tongue usually provided on each side with a row of conspicuous "gill rakers"; gill openings wide.

In most seas, many species entering fresh water to spawn, a few remaining permanently. As here restricted, there are about twelve genera, and one hundred and twenty species. Many are highly valued as food fishes.

* Upper jaw not projecting beyond the lower. (CLUPEINÆ.)

† Teeth wanting or on tongue only (rarely a few weak teeth in jaws) no dorsal filament; scales regularly arranged, not ciliated; upper jaw emarginate.

‡ Depth 3⅓ in length; cheeks higher than long; preopercle with a very short horizontal process. . ALOSA, 1.

‡‡ Depth 3⅔ to 3¾ in length; cheeks longer than high; preopercle with an oblong horizontal process. POMOLOBUS, 2.

** Upper jaw projecting beyond the lower. (DOROSOMINÆ.)

a. Body compressed, deep; last ray of dorsal filamentous; mouth toothless. . . . . . . DOROSOMA, 3.

1. *ALOSA*, Cuvier. SHADS.

1. ***A. sapidissima,*** (Wilson) Storer. COMMON SHAD. Head 4¼ in length; eye 5 in head; bluish, sides silvery; scales large; D. 18; A. 21; V. 9; lat. l. 68. Newfoundland to Florida, entering rivers; also lately introduced into Western streams; a valuable food fish. (*A. præstabilis*, DeK.)

2. *POMOLOBUS*, Rafinesque. GASPEREAUS.

1. ***P. pseudoharengus,*** (Wils.) Gill. ALEWIFE. GASPEREAU. SPRING HERRING. Head 5 in length; eye 4 in head; bluish, sides iridescent; D. 18; A. 18; V. 9. Newfoundland to Florida, entering rivers, sometimes land-locked in ponds; a common food fish. (*A. tyrannus*, DeK. *A. cyanonoton*, Stor., etc., etc.)

Var. ***lacustris,*** Jordan. CAYUGA LAKE SHAD. Head 4 in length; body much heavier forward than in the others; depth of head 4¼ in length of body; eye large, longer than snout, 3 in head; scales large, loose; caudal peduncle in its narrowest place not half wider than eye; steel blue, punctate; sides silvery; D. 15; A. 19; lat. l.

45; 33 scutes in all, 13 behind ventrals. Cayuga L., N. Y., dredged in deep water. (*Wilder.*)

2. ***P. chrysochloris,*** Raf. OHIO GOLDEN SHAD. SKIP JACK. Head 4 in length; eye $4\frac{1}{4}$ in head; body elliptical, much compressed; scales large, high, rather firm; depth of head $5\frac{1}{2}$ in length of body; caudal peduncle about twice width of eye; brilliant blue with green and golden reflections, silvery below; D. 18; A. 18; lat. l. 55; 17 scutes behind ventrals. Ohio R. and lower Mississippi; a handsome species.

3. DOROSOMA, Rafinesque. GIZZARD SHADS.

= *Chatoessus*, Cuvier.

1. ***D. cepedianum,*** (Lac.) Gill. HICKORY SHAD. GIZZARD SHAD. Head 4 in length; depth $2\frac{3}{5}$; origin of dorsals behind ventrals, nearer snout than caudal; uniform bluish gray; often with a dark shoulder blotch; D. 13; A. 32; lat. l. 55. Cape Cod to Cape Hatteras, chiefly marine, but often land-locked in ponds, where it becomes *D. insociabile*, Abbott.

2. ***D. notatum,*** Raf. OHIO GIZZARD SHAD. THREAD SHAD. Head $3\frac{1}{3}$ in length; depth $2\frac{3}{4}$ to 3 in length; dorsal about midway, slightly behind ventrals; dorsal filament nearly one-fourth length of body; bluish, sides bright silvery. Ohio R. and lower Mississippi, apparently not descending to the sea; not well distinguished from the preceding and perhaps the same. (*C. ellipticus*, Kirt.)

SUB-ORDER.—EVENTOGNATHI.

(*The Carp-like Fishes.*)

FAMILY CVIII.—CYPRINIDÆ.

(*The Minnows.*)

Head naked, body scaly (except in *Meda*, etc.); margin of upper jaw formed by intermaxillaries alone; mouth toothless; lips much less developed than in the *Catostomoids;* barbels two to four (absent in most of our genera and not large in any); lower pharyngeal bones well developed, falciform, nearly parallel with the gill arches, each provided with one to three series of teeth in small number, never more than seven; belly usually rounded, rarely compressed, never serrated; gill openings moderate, separated by a narrow isthmus; no adipose fin; dorsal fin (in all our species) short, of less than ten rays; air bladder usually large, commonly divided into an anterior and a posterior lobe, rarely wanting; stomach without appendages, appearing as a simple enlargement of the intestines.

Small fishes of the fresh waters of the Old World and of North America. Genera about one hundred and fifty, species seven hundred to one thousand; excessively abundant where found, both in individuals and in species, and from their great uniformity in size, form and coloration, constituting one of the most difficult groups in all Natural History in which to distinguish species. Ours are mostly of smaller size than those of the Old World, several of the larger European types being represented in America by *Catostomoid* forms. Our largest species, *Semotilus rhotheus*, rarely attains a weight of three or four pounds, and a length of nearly eighteen inches.

The smaller *Hybopses* and *Hemitremiæ* scarcely reach a length of two inches.

The spring or breeding dress in many genera is peculiar. Often the top of the head, and sometimes the whole dorsal region also, is covered in the males with rows of spinous tubercles, outgrowths from the epidermis, and usually the skin of the muzzle is then swollen and charged with pigment. In *Semotilus* and *Ceratichthys* these tubercles are quite large and cover the front and sides of the head; in *Pimephales* and *Hyborhynchus* they are placed entirely on the front of the obtuse snout; in *Campostoma* the whole dorsal region, and sometimes the whole body, is rough with large tubercles; in *Luxilus*, *Plargyrus*, *Lythrurus*, *Gila* and *Minnilus* the prickles are quite small and crowded on the upper surface of the head and neck.

In some genera, the males in spring are adorned with bright tints of red, which give these little fishes a temporary brilliancy scarcely surpassed even by Trouts or Darters. In *Luxilus*, *Lythrurus*, *Campostoma*, and *Semotilus*, the red appears chiefly as pigment in the membranes of some or all of the fins, the sides of the body being usually more or less flushed; in *Rhinichthys* and *Gila*, the black of a portion of the lateral band usually changes to red; in *Chrosomus*, and probably *Phoxinus*, the pigment lies mostly in the skin of the belly, and in *Minnilus* it is chiefly about the head and the bases of the fins. In *Pimephales* and *Hyborhynchus*, black pigment is deposited in the skin of the head, and in the species of the sub-genus *Plargyrus*, satin-white pigment occurs in the fins. So far as is known to me, the species of *Hybopsis*, *Hemitremia*, *Hybognathus*, *Photogenis*, *Phenacobius*, *Exoglossum*, *Notemigonus*, *Ericymba*. and the sub-genus *Erinemus* of *Cera-*

tichthys, with one or two exceptions, show no special variations in the breeding season, but this matter needs further investigation.

The genera given below appear to be well characterized, although several are very closely related, and the occurrence of intermediate forms may require them to be reunited. *Lythrurus* and *Plargyrus* are properly sub-genera of *Luxilus*, but they show so many external peculiarities that I have, for the present, given them generic rank. Most of the species here admitted have been pretty thoroughly tested by the author, though there are several, particularly under *Hybopsis*, which I do not like to indorse.

NOTE.--Young *Cyprinidæ* usually are more slender than adults of the same species, and the eye is always much larger; they also frequently show a black lateral stripe and caudal spot which the adults may not possess. Spots on the fins are generally characteristic. The following artificial key will generally hold good for *adult* fishes, but only great patience and long and careful observation will enable the student to identify the young. The accounts of the pharyngeal teeth are taken from Prof. *Cope's* invaluable "Monograph of the *Cyprinidæ* of Pennsylvania," but most of them have been verified by the author. It has been thought best to make the dental characters subordinate in the present work, but the student is strongly advised to examine the teeth of these fishes, as the actual characters of the genera are largely drawn from them.

* Native species; fins without serrated spines.

† Dorsal preceded by a short, spinous ray, which is connected by a membrane to the soft rays (about half the height of the fin, and appearing as if broken off); snout short and blunt, overlapping the small mouth; front of muzzle with about a dozen large tubercles in spring males; fins low; dorsal with a black spot in front, about half way up; peritoneum black; intestines long, two to three times length of body; teeth one-rowed, 4-4, with masticatory surface.

a. Lateral line complete; body elongated; angle of mouth usually with a small barbel. . HYBORHYNCHUS, 4.

aa. Lateral line incomplete; body very short; no barbels. PIMEPHALES, 3.

†† Dorsal without spine, the rudimentary ray in front smaller and firmly attached to the first developed ray.

b. With a small barbel (often very minute) at each angle of the mouth; alimentary canal not longer than body; teeth hooked, without masticatory surface.

c. Intermaxillaries not projectile, skin of lip and front continuous; mouth rather inferior, beneath the prominent snout; scales small, usually mottled; lat. l. 60 to 70; sides rosy in Spring; isthmus wide; teeth 2, 4–4, 2. RHINICHTHYS, 9.

cc. Intermaxillaries projectile; species usually of larger size or with larger scales.

d. Mouth very wide, oblique; head broad, rounded; jaws nearly equal; the minute barbel just above the angle of the mouth; often a black spot on front base of dorsal; teeth 2, 5–4, 2. . . . . SEMOTILUS, 7.

dd. Species usually of smaller size; the mouth rather narrow and more or less inferior, with the small but evident barbel at its angle; no spot at base of dorsal; teeth 4–4; or 1 or 2, 4–4, 1 or 2. CERATICHTHYS, 8.

bb. Angle of mouth without traces of a barbel.

e. Mouth terminal, more or less oblique; the jaws about even, or the lower somewhat projecting. (Upper jaw sometimes swollen and projecting in spring males of *Luxilus* and *Semotilus*.)

f. Anal with twelve or more developed rays; body elevated, abdomen compressed; lateral line greatly decurved; mouth short, oblique; lower jaw rather longest; dorsal behind ventrals; intestinal canal long; teeth one-rowed, crenate, with masticatory surface; rather large species, with a silvery or golden lustre. . . . . . . NOTEMIGONUS, 21.

ff. Anal with seven to eleven (rarely twelve) developed rays.

g. Lateral line incomplete or wanting; mouth oblique; size small.

h. Scales very small; lat. l. more than 60; dorsal behind ventrals.

i. Body very stout; head short and heavy; mouth large; a single black lateral band; back unspotted; alimentary canal short; teeth two-rowed, inner 5–4, without masticatory surface. PHOXINUS, 14.

ii. Body moderately stout, graceful; sides with one or two black bands; back spotted; belly, etc., brilliant red in spring; alimentary canal elongate; teeth one-rowed, 4–5 or 5–5, with masticatory surface. . CHROSOMUS, 13.

hh. Scales rather large; lateral line less than 40; small species with a dark lateral band; teeth 4–4 or 5–4, with masticatory surface. HEMITREMIA, 12.

gg. Lateral line complete (rarely obsolete on the last four or five scales); alimentary canal short.

j. Dorsal decidedly behind ventrals; mouth oblique, usually large, the lower jaw commonly projecting; elongated species, generally of small size, more or less compressed, with the caudal peduncle long; males rosy and with small tubercles in spring; teeth two-rowed.

k. Scales very small; lat. l. 45 to 75; dorsal without spots; scarcely silvery; sides and below red in spring; anal fin short, with eight or nine rays; teeth 5–4 (inner series), without masticatory surface. . . . . GILA, 15.

kk. Scales rather small, especially in front of dorsal, larger along the sides and crowded so that the exposed surfaces are decidedly higher than long as in *Luxilus;* lat. l. 40 to 50; fins very high, bright red in the spring, a large black spot at base of dorsal in front; anal fin long, of ten or eleven rays; teeth 2, 4–4, 2, with masticatory surface. . LYTHRURUS, 16.

kkk. Scales rather large, not closely imbricated, brilliantly silvery; lat. l. 35 to 40; fins unspotted, uncolored; forehead, etc., rosy in spring: anal fin rather long, of nine to twelve rays; body elongate; teeth (inner series) 4–4, usually without masticatory surface.

MINNILUS, 20.

jj. Dorsal directly above ventrals (rarely slightly posterior); anal short, of eight or nine rays.

l. Lateral line with less than 45 scales.

m. Scales quite large, closely imbricated, the exposed portion being much higher than long, especially on the sides of the body, very conspicuously so in adults; body compressed; lateral line decurved; mouth oblique; males with the head tuberculate in spring.

— Dorsal fin directly over ventrals, without distinct black spot; sides and fins with red pigment in spring; mouth and eyes large; anal commonly I, 9; teeth 2, 4–4, 2 with masticatory surface.

LUXILUS, 17.

— Dorsal fin slightly behind ventrals, with a large black blotch behind; pigment on sides and fins pure satin white; mouth and eyes rather small; A. normally I, 8; teeth 1, 4–4, 1, sometimes crenate. . . PLARGYRUS, 18.

mm. Scales normal, not closely imbricated.

n. Elongated, compressed species with the mouth very oblique, and the lower jaw usually projecting; scales generally brilliantly silvery; eye large; lateral line 35 to 40; inner row of teeth 4–4, without masticatory surface. . PHOTOGENIS, 19.

nn. Small weak species, with the head short, mouth small, and the lower jaw usually not projecting; scales scarcely silvery; inner row of teeth 4–4, with masticatory surface. . . . HYBOPSIS, 11.

ll. Lateral line with more than 45 scales; head broad and large; body little compressed; teeth without masticatory surface, 2, 5–4, 2. SEMOTILUS, 7.

ee. Mouth inferior, scarcely oblique, the upper jaw being notably longest.

1. Intestinal canal about eight times the length of the body, its numerous convolutions entirely surrounding the small air-bladder; peritoneum black; head rather long and narrow, with sub-vertical cheeks; lips with cartilaginous sheaths; eyes well back and high up the head, the iris orange-colored; scales rather small, more or less mottled; a blackish vertical bar behind head and a dusky band across dorsal and anal, which in spring males is bordered with fiery orange; back and head coarsely tuberculated in ♂ in spring; teeth 4–4, with oblique masticatory surface. . . . . . . CAMPOSTOMA, 2.

2. Sub-orbital bone, interopercle and base of mandible much dilated, cavernous, crossed by large mucous channels (these readily seen under any circumstances, by looking at the head of the fish from *below*); snout thick; slender fishes, brightly silvery; teeth 1, 4–4, 0, hooked, without masticatory surface. ERICYMBA, 6.

3. Mandible much contracted, with a lobe on each side at base; the middle portion appearing like a protruding tongue; intermaxillaries not projectile; stout species, dusky in color; teeth 1, 4–4, 1, hooked, without masticatory surface. . . . EXOGLOSSUM, 1.

4. Lips large, defended by a cartilaginous sheath on their opposing edges, somewhat plicate or tuberculate;

elongate species of small size, resembling young suckers; teeth 4–4, hooked, sharp-edged.
PHENACOBIUS, 10.

5. Intermaxillaries not projectile, the skin of lip and forehead being continous; small dark species with 60 or more scales in the lateral line.
RHINICHTHYS, 9.

6. Scales high, closely imbricated; body compressed, silvery; dorsal with a black blotch behind; snout with small tubercles and fins with satiny pigment in spring; teeth 1, 4–4, 1. . . PLARGYRUS, 18.

7. Small fishes — silvery or plumbeous, with none of the preceding combinations.

o. Jaws with sharp cutting edges; intestines much convoluted, about four times the length of the head and body; peritoneum black; size moderate; scales brightly silvery; teeth 4–4, with oblique masticatory surface and no hook. . HYBOGNATHUS, 5.

oo. Jaws not trenchant; intestines not longer than head and body; peritoneum pale; scales various, rather large; small, weak species; teeth one or two-rowed, inner series 4–4. . . HYBOPSIS, 11.

** Introduced species; dorsal very long and anal short, each being preceded by a stout spine which is serrated behind.

p. Mouth with four long barbels; teeth molar 3, 1–1, 3.
CYPRINUS, 23.

pp. No barbels; teeth compressed, 4–4. . CARASSIUS, 22.

1. EXOGLOSSUM, Rafinesque. STONE TOTERS.

1. ***E. maxillingua,*** (LeS.) Haldeman. DAY CHUB. CUT-LIPS. NIGGER CHUB. Body stout; depth 4¼ in length, head 4; eye small, nearly 5 in head; dorsal behind midway between snout and caudal; dusky above, a blackish shade along caudal peduncle; D. I, 8; A. I, 8; lat. l. 50 to 55; L. 4 to 6. W. N. Y. (Susquehanna basin)

to Tenn. and S.; a fish of remarkable appearance, singularly distinguished from all our other *Cyprinidæ* by the three-lobed lower jaw.

2. ***E. mirabile,*** Grd. Western Stone Toter. Head 5 in length; dorsal nearer snout; D. I, 9; A. I, 8. Arkansas R.

2. CAMPOSTOMA, Agassiz. Stone Luggers.

1. ***C. anomalum,*** (Raf.) Ag. Stone Lugger. Stone Roller. Brownish, with a brassy lustre above, the scales more or less mottled with dark; a black vertical bar behind opercle; iris usually orange-red; dorsal and anal each with dusky cross-bar about half way up, the rest of the fin olivaceous, or in spring males fiery orange; males in spring with many rounded tubercles on head, and usually the whole upper surface—in no other genus are these nuptial appendages so extensively developed—scales deep, rather small and crowded anteriorly; intestinal canal six to nine times the total length of the body, its numerous convolutions passing above and *around* the air-bladder, an arrangement found in *Campostoma* alone among all the Vertebrates; D. I, 8; A. I, 7; lat. l. 50 to 55; L. 4 to 8; herbivorous. Mississippi Valley, everywhere abundant; one of the most curious and interesting of American fishes. [*C. dubium*, (Kirt.) Cope. *C. callipteryx*, *gobioninum*, etc., Cope.]

3. PIMEPHALES, Rafinesque. Round-Headed Minnows.

1. ***P. promelas,*** Raf. Fat-Head. Black Head. Head almost globular, black in adult males; snout in ♂ with several large tubercles; body very short and deep; scales crowded; eye small; mouth very small and short; a large black dorsal blotch; males dusky; females oliva-

ceous; D. I, 7; A. I, 7; lat. l. 46; L. 2½. Ohio Valley to Upper Missouri. Known at sight, as it resembles nothing else.

2. ***P. milesii,*** Cope. Miles' Minnow. Snout longer; eye larger; mouth larger; color paler, usually a blackish lateral stripe; D. I, 8; A. I, 7; lat. l. 40. Mich. to Ky. (*P. agassizii*, Cope.)

4. *HYBORHYNCHUS,* Agassiz. Blunt-Nosed Minnows.

1. ***H. notatus,*** (Raf.) Ag. Blunt-Nosed Minnow. Brownish or bluish, a dusky shade along sides sometimes forming a caudal spot; a distinct black spot on middle of front rays of dorsal; head short; snout in spring males with disproportionately large tubercles, usually fourteen in all; a distinct barbel at each angle of the mouth; scales in front of dorsal small and crowded; D. I, 8; A. I, 7; lat. l. 45; L. 3 to 4. N. Y. to Tenn., Wis., and Mo.; very abundant in the Ohio Valley. (*H. superciliosus*, Cope. This form, said to be distinguished from the true "*notatus*" by the presence of the barbel, is the only one I have yet seen. Specimens from Rafinesque's original locality possess the barbel.)

5. *HYBOGNATHUS,* Agassiz. Blunt-Jawed Minnows.

1. ***H. nuchalis,*** Ag. Blunt-Jawed Minnow. Smaller and more dusky than the next; eye small, shorter than snout, 4 to 4½ in head; depth 4½ in length, about equal to length of head; scales in front of dorsal very small and crowded; D. I, 8; A. I, 8; lat. l. 38; L. 2½. Ohio Valley and W. This and the next may be readily known from the *Hybopses*, which they strongly resemble externally, by the peculiarities of the intestines.

2. ***H. argyritis,*** Grd. Silvery Minnow. Olivaceous

green above, sides clear silvery with bright reflections; fins unspotted; eye large, longer than muzzle, 3 to 4 in head; depth 4⅛ in length; scales in front of dorsal quite large; lateral line decurved; head large, upper jaw heavy; D. I, 8; A. I, 8; lat. l. 38; L. 5. N. J. to N. C. and W. to the Upper Missouri; abundant in the larger streams; one of our handsomest dace. (*H. osmerinus*, Cope, not in the least different!)

6. ERICYMBA, Cope. Ericymbas.

1. ***E. buccata,*** Cope. Silver-Mouthed Dace. Elongated; depth nearly 5 in length; head 4; eye large, 3 in head; olivaceous above, sides brilliantly silvery, a narrow vertebral line, and a lateral chain of brown dots; upper jaw rather large, its profile angulated; mucous channels in lower jaw very conspicuous; D. I, 8; A. I, 8; lat. l. 33; L. 5. Ohio Valley, Penn. to Tenn. (*Jordan*) and Illinois, abundant. A beautiful little fish singularly distinguished from all our other species by the cavernous bones of the head.

7. SEMOTILUS, Rafinesque. Fall Fish.

* Scales moderate, crowded forwards, 55 or more in lateral line; a black spot at base of dorsal in front. (*Semotilus.*)

1. ***S. corporalis,*** (Mitch.) Putnam. Common Chub. Horned Dace. Body stout, depth 4½ in length; head large, 3¾; dusky above, especially along edges of scales; sides bluish, a black lateral band in young; silvery below, sides and fins flushed with crimson in spring; D. I, 8; A. I, 8; lat. l. 55 to 65; L. 10 to 12. New England (Housatonic R., *Jordan*) to the Missouri region and S.; the most widely diffused of our *Cyprinidæ*, excepting *Ceratichthys melanotus*. It may be known under all circumstances by the large head and the

peculiar dorsal spot. (*S. atromaculatus*, *dorsalis*, *cephalus*, *speciosus*, etc., etc., of authors.)

Var. ***pallidus,*** (Grd.) Jordan. Pale Chub. Differs in its pale color and slightly smaller scales. Missouri region and S.

** Scales larger, scarcely crowded anteriorly; lat. l. 45 to 55; no dorsal spot. (*Leucosomus*, Hæckel.)

2. ***S. argenteus,*** (Storer) Putn. Eastern Chub. Roach. Brownish, sides roseate; depth 4½ in length; head 4; eye 5 in head; D. I, 8; A. I, 8; lat. l. 51; L. 12 to 14. New England and New York. [*S. pulchellus*, (Storer) Gill.]

3. ***S. rhotheus,*** Cope. Big Chub. Rosy Fall Fish. Steel blue above, sides silvery, rosy in spring; proportions of the last; D. I, 8; A. I, 8; lat. l. 46; the largest of our *Cyprinidæ*, reaching a length of 18 inches. Mass. to Md., in the larger streams.

8. *CERATICHTHYS,* Baird. Horned Chubs.

> *Nocomis*, Grd.

< *Gobio*, Cuvier (European).

* Mouth nearly terminal; rather large species, scarcely silvery, resembling *Semotilus*. (*Ceratichthys*.)

† Lateral line with 40 to 45 scales.

1. ***C. melanotus,*** (Raf.) Jordan. Horned Chub. Jerker. Bluish olive, sides with bright green and coppery reflections; a curved blotch behind the opercle; fins pale orange, unspotted; white below, rosy in spring; adult males in the spring with the top of the head very much swollen, elevated into a sort of crest, sometimes nearly one-third of an inch higher than the level of the neck, covered with large tubercles; a stout species, with large scales which are not crowded

anteriorly; young with a dark caudal spot; head 4 in length; depth nearly the same; D. I, 8; A. I, 7; lat. l. 40 to 45; L. 6 to 9. Penn. to Utah and S.; abundant almost every where; the most widely diffused of all our fresh water fishes. [*C. biguttatus*, (Kirt.) Bd., *C. stigmaticus*, *cyclotis*, etc., Cope.]

†† Lateral line about 60.

2. ***C. plumbeus,*** (Ag.) Gthr. LEAD-COLORED CHUB. Depth = length of head, 4 to 5 in body; mouth small; dusky; size large; D. I, 8; A. I, 7; lat. l. 60. L. Superior. (*C. prosthemius*, Cope.)

** Mouth small, inferior — upper jaw notably longest; barbels rather long; small, silvery species, resembling *Hybopsis*. (*Erinemus*, Jordan.)

‡ Lateral line 30 to 42.

3. ***C. hyalinus,*** Cope. BIG-EYED MINNOW. Olivaceous or bluish, sides clear silvery; eyes very large, 3 in head; depth 5 in length; head rather large, 4; D. I, 9; A. I, 8; lat. l. 40; L. 3. Ohio Valley, abundant; the smallest species, resembling *Photogenis arriommus*, but with a very different mouth.

Var. ***labrosus,*** (Cope) Jordan. LARGE-LIPPED MINNOW. Similar, but with the scales larger, the body slimmer, and the barbels and lips more developed; D. I, 8; A. I, 8; lat. l. 34. Rivers of N. C. and Tenn.

‡‡ Lateral line 45 to 50; long, slender species, with the snout projecting.

4. ***C. dissimilis,*** (Kirt.) Cope. SPOTTED SHINER. Pale olivaceous, sides bright silvery, with a bluish lateral band, widened at intervals into spots; fins immaculate; depth 5 in length; head 4; eye large, 3½ in head; D. I, 8; A. I, 7; lat. l. 47 to 50; L. 6. Ohio Valley and Lake region, not uncommon.

5. ***C. monachus,*** Cope. SOLITARY CHUB. Similar, a black spot on last rays of dorsal; a dark caudal spot; eye small, 4 in head; lat. l. 56. Holston R.

‡‡‡ Lateral line about 70.

6. ***C. cataractæ,*** (Val.) Cope. NIAGARA GUDGEON. Slender, depth 6 in length; head 4; D. I, 8; A. I, 8; lat. l. 70. Niagara Falls.

9. *RHINICHTHYS,* Agassiz. LONG-NOSED DACE.

= *Argyreus*, Hæckel (preoccupied).

* Snout projecting considerably beyond the mouth; body slender, depth usua.ly 5 to 6 in length; barbels evident.

1. ***R. nasutus,*** (Ayres) Ag. LONG-NOSED DACE. Brownish, mottled, not banded; eye half the length of the long snout; head $3\frac{4}{5}$ in length; D. I, 8; A. I, 7; lat. l. 63; L. 5. New England to Va. and Wis., in clear brooks.

2. ***R. marmoratus,*** Ag. MARBLED DACE. Brown, marbled; eye $2\frac{1}{2}$ in snout; head $4\frac{1}{4}$ in length; lat. l. 70. Great Lakes.

** Snout scarcely projecting; body stout; depth 4 to 5 in length; barbels scarcely visible.

† A distinct dark band from snout to caudal (red in spring.)

3. ***R. atronasus,*** (Mitch.) Ag. BLACK-NOSED DACE. Dusky, belly silvery; lateral band bright crimson in spring, becoming orange in summer, black at other times; fins often rosy in spring; depth $4\frac{2}{3}$ in length; head $3\frac{3}{4}$; D. I, 8; A. I, 7; lat. l. 65. New England to Ohio Valley, in clear brooks; abundant Eastward.

4. ***R. obtusus,*** Ag. BROWN-NOSED DACE. Similar; sides with a brown band, edged above and below with paler; head 4 or more in length; D. I, 8; A. I, 8; lat. l. 63 to 70. Western streams.

†† Sides without band, or with merely a dusky shade.

5. ***R. lunatus,*** Cope. FORK-TAILED DACE. Reddish brown, with dusky spots; depth 4⅓ in length; head 4; lat. l. 60. Western streams.

10. ***PHENACOBIUS,*** Cope. PHENACOBIES.

1. ***P. teretulus,*** Cope. Head=depth, 4¾ in length; lips plicate; a dusky lateral band; D. I, 8; A. I, 7; lat. l. 43. Streams of W. Va.

2. ***P. uranops,*** Cope. Head 4¾ in length; depth 6¼; lips tuberculate; lat. l. 60. Holston R.

11. ***HYBOPSIS,*** Agassiz. BLUNT-FACED MINNOWS.

* Mouth inferior, horizontal, small.

† Head 5¼ to 6 times in total length, including caudal fin. (*Hudsonius*, Grd.)

1. ***H. storerianus,*** (Kirt.) Ag. STORER'S MINNOW. Silvery, scales with black dots, forming a dark lateral stripe; snout blunt, about as long as eye; depth 5½ in length; D. I, 9; A. I, 9; lat. l. 41. Great Lake region, etc.

2. ***H. hudsonius,*** (Clinton) Putnam. SPAWN-EATER. Silvery, often with dark shades; snout much shorter than eye; depth 4 in length; D. I, 8; A. I, 8; lat. l. 38. Lakes and rivers; abundant eastward. (*Huds. fluviatilis* and *amarus*, Grd. *H. phaënna*, Cope.)

†† Head 4½ to 5 in length, inclusive of caudal fin. (*Hybopsis.*)

‡ Pectoral fins short, not reaching ventrals.

a. Lateral line 43 to 45.

3. ***H. tuditanus,*** Cope. Form and coloration of *Hyborhynchus notatus*, but said to want the dorsal spine; head broad and blunt. L. Michigan; "a doubtful species," as are several others in this genus.

aa. Lat. l. 36 or 37.

4. ***H. spectrunculus,*** Cope. Eye large; head broad and flat; a plumbeous lateral band and black caudal spot; fins reddish; A. I, 9. Holston R.

5. ***H. stramineus,*** Cope. Silvery; head more rounded; body plump; scales in front of dorsal 12 to 14, quite large. Mich. to Ind., abundant.

aaa. Lat. l. 32 or 33.

6. ***H. procne,*** Cope. Very similar, but the caudal peduncle contracted and slender, scales large; a plumbeous band over black pigment. D. I, 8; A. I, 7. Penn., etc., common E.

‡‡ Pectorals elongated, nearly or quite reaching ventrals.

7. ***H. microstomus,*** (Raf.) Jordan. LONG-HEADED MINNOW. Head elongated; a silvery band along sides and a series of black dots along lateral line; depth 5 in length; caudal peduncle not abruptly contracted; pectorals long, about reaching ventrals; D. I, 8; A. I, 7; lat. l. 33. Va. to Ky. [*H. gracilis*, Ag. (type of genus.) *H. longiceps*, Cope.]

8. ***H. volucellus,*** Cope. Elongated; head long; a dusky lateral band; D. I, 8; A. I, 7; lat. l. 34. Detroit R., (same as preceding?)

** Mouth larger, oblique, the lower jaw about as long as upper. (*Alburnops*, Grd.)

b. Lateral line 35 or 36, a dark lateral band.

9. ***H. chalybœus,*** Cope. PIGMY MINNOW. Muzzle flat; head 4 in length; caudal peduncle abruptly slender, lateral band very distinct, shining black; A. I, 8. Penn., N. J.; one of the smallest of the *Cyprinidæ;* length 1½ inches; (resembles *Hemitremia bifrenata.*)

10. ***H. fretensis,*** Cope. Silvery, a plumbeous lateral band; A. I, 8. Detroit R.; resembles a *Minnilus.*

bb. Lateral line 38 to 39.

11. ***H. plumbeolus,*** Cope. LEAD-COLORED MINNOW. Uniform leaden silvery; body compressed; lower jaw projecting; A. I, 9. Great Lakes.

12. ***H. regius,*** (Girard) Cope. SILVERY MINNOW. Much elongated, compressed; uniform silvery; D. I, 10; A. I, 9; lat. l. 38. Potomac R.

13. ***H. rubricroceus,*** Cope. RED-BANDED MINNOW. A red lateral band and red touches on head; belly yellow; A. I, 9; lat. l. 38. Upper Tennessee; a brightly colored species.

bbb. Lateral line 42 to 45; colors dull.

14. ***H. hæmaturus,*** Cope. RED-TAILED MINNOW. A black spot on dorsal and at base of caudal; dusky; tail brick-red; A. I, 7. Tributaries of Lake Michigan.

12. HEMITREMIA, Cope. HEMITREMES.

1. ***H. vittata,*** Cope. SOUTHERN HEMITREMIA. Dusky; a black lateral band, and above this several paler and smaller ones, the upper running into the dorsal line; fins small; depth 4 in length; head 4½; D. I, 8; A. I, 7; lat. l. 34; L. 2. Head waters of Tennessee and Cumberland Rivers. (Description from Kentucky specimens.)

2. ***H. heterodon,*** Cope. NORTHERN HEMITREMIA. Head = depth, about 4 in length; snout flat, rather pointed; back compressed, elevated; olive, a dusky lateral shade; D. I, 8; A. I, 8; lat. l. 35. Mich., Wis.

3. ***H. bifrenata,*** Cope. EASTERN HEMITREMIA. Head = depth, 4⅕ in length; snout blunt; olive, a burnished, jet-black lateral band of a deeper color than in any other small minnow; D. I, 8; A. I, 7; lat. l. 36. Mass. to Md., abundant.

13. CHROSOMUS, Rafinesque. RED-BELLIED MINNOWS.

1. ***C. pyrrhogaster,*** Jordan. CHROSOMUS. RED-BELLIED DACE. Brownish olive, with black spots on the back, a black or brown band from above the eye, straight to the tail; another below, running through eye, decurved along the lateral line; belly and space between bands bright silvery, brilliant scarlet red in spring males, as are the bases of the vertical fins; a dark vertebral line; females obscurely marked; D. I, 8; A. I, 9; lat. l. 80 to 90. Penn. to Wis. and S., abundant in small streams; one of the most beautiful of our fishes. [*C. erythrogaster*, (Kirt.) Ag., not of Raf.]

2. ***C. erythrogaster,*** Raf. KENTUCKY RED-BELLY. Similar but even more brilliant; a dark band through eye to base of anal; another from the middle of the body to caudal; above this, back with distinct cross-bars and spots; belly silvery or crimson; fins orange and yellow; lat. l. 67. Ky. to Va. and N. C. (*C. oreas*, Cope.)

3. ***C. eos,*** Cope. Lateral bands confluent on caudal peduncle; teeth 5–5; lat. l. 75. Susquehanna R.

14. PHOXINUS, Rafinesque (1820!). EUROPEAN MINNOWS.

1. ***P. neogæus,*** Cope. NEW WORLD MINNOW. Blackish above, a broad black lateral band through eye, becoming a spot on the tail; belly white; fins dusky; head large, 3½ in length; depth rather less; mouth large, oblique; eye large; D. I, 7; A. I, 7; L. 3. Southern Mich. (*Cope*); Baraboo R., Wis. (*Bundy*); a curious fish, related to the Minnow of Europe (*P. lævis*, Ag.)

15. GILA, Baird and Girard. Leather-Sided Minnows.

* Scales very small; mouth large, very oblique, the lower jaw projecting. (*Clinostomus*, Grd.)

1. ***G. elongata,*** (Kirt.) Jordan. Red-Sided Minnow. Dark bluish, mottled by paler scales; sides with a broad black band, the front half of which is bright crimson in the spring; a dark dorsal stripe; mouth very large, the lower jaw narrowed and projecting farther than in any other of our Dace; a little knob at the tip which overlaps the end of the upper jaw; body much elongated, but little compressed; depth 5 in length; head 4¼; eye moderate, about 3½ in head; D. I, 8; A. I, 8; lat. l. 65 to 70; L. 4. Great Lakes, Ohio Valley, etc.; a handsome species. (*C. proriger*, Cope.)

2. ***G. vandoisula,*** (Val.) Jordan. Rosy Dace. A light and a dark lateral band; snout pointed; mandible shorter than in the preceding; eye larger; depth 5 in length; head 4; D. I, 9; A. I, 8; lat. l. 48. Streams about Chesapeake Bay and S. (*C. funduloides*, Grd.)

3. ***G. affinis,*** (Grd.) Cope. Body deeper, depth = length of head, 3⅔ in body; eye rather small; D. I, 9; A. I, 8; lat. l. 46. James R. (*C. carolinus*, Grd.)

4. ***G. margarita,*** Cope. No lateral band; snout obtuse; depth 5 in length; head 4; lat. l. 58. Penn.

16. LYTHRURUS, Jordan. Red-Fins.

< *Hypsilepis*, Cope.

1. ***L. diplœmius,*** (Raf.) Jordan. Red-Fin. Bright steel blue, with purplish shades, silvery below; a large black spot on the anterior rays of dorsal in front; fins otherwise unicolor, plain olivaceous in ♀, brilliant brick red in spring males; scales with more or less dark edging; nuptial tubercles minute, very numerous, whitish,

chiefly on the upper surface of head; body much compressed; back elevated; head deep, rather obtuse; depth $3\frac{3}{4}$ in length; D. I, 9; A. I, 10; lat. l. 44; L. 3. Western streams, generally abundant; an exceedingly brilliant fish in the breeding season; known at all times by the dorsal spot and compressed body, with large fins and long caudal peduncle. (*Rutilus ruber*, Raf.) (Not *Leuciscus diplemius*, Kirt.)

2. ***L. ardens,*** (Cope) Jordan. SOUTHERN RED-FIN. Colors similar, but red on sides more conspicuous; head rather pointed, with the mouth still more oblique; depth 5 in length; D. I, 9; A. I, 11; lat. l. 50. Cumberland and Roanoke Rivers and S.

17. LUXILUS, Rafinesque. SHINERS.

= *Hypsilepis*, Baird.

* Fins and lower parts with rose-red pigment in spring and summer; no distinct black dorsal spot; eye large; mouth large, oblique, the lower jaw about as long as upper in closed mouth; species of large size; "Red Fins."

1. ***L. cornutus,*** (Mitch.) Jordan. COMMON SHINER. RED-FINNED SHINER. ROUGH-HEAD. RED-FIN. Adult deep steel blue or olivaceous above, with golden vertebral and lateral bands, very conspicuous in life; sides silvery, rosy in males in spring; fins plain olivaceous or somewhat dusky, becoming crimson in spring; young olivaceous and silvery, not closely resembling the adult; depth 3 to 5 in length, greater than length of head in adults; head large; mouth moderately oblique, the lower jaw not projecting; lateral line much decurved; D. I, 8; A. I, 9; lat. l. 40 to 45; L. 6. U. S. from Maine to the Rocky Mountains, every where abundant, and extremely variable. The adults may be known at once by the high and narrow exposed surfaces of the scales;

the young often need close attention. (*Plargyrus typicus*, Grd.; *L. chrysocephalus*, Raf.; *L. diplemius* and *plargyrus*, Kirt.; *Leuciscus frontalis*, Ag., a stout variety from the Great Lakes.)

2. ***L. coccogenis,*** (Cope) Jordan. RED-CHEEKED SHINER. Steel blue, sides silvery; dorsal, caudal, and pectorals red in the male; adults of both sexes with the upper jaw, base of dorsal, and a vertical streak down the cheeks, bright orange red, the latter mark appearing like a brand; dorsal and caudal with a broad dusky bar; slimmer than *cornutus*; depth 4⅓ in length, about equal to length of head; mouth very oblique, the lower jaw projecting (excepting in tuberculate males); D. I, 7; A. I, 9; lat. l. 42. Head waters of the Tennessee R., abundant; a beautiful and very distinct species.

18. PLARGYRUS, Rafinesque. SILVER FINS.

< *Rutilus*, Raf. (= *Leuciscus*, Klein, European.)

< *Hypsilepis*, Cope.

* Fins and lower parts with milk-white pigment in spring, never red; dorsal with a large black spot on the last rays, about half way up (an important feature); upper jaw projecting in closed mouth; eye rather small; species of rather small size; "Silver-Fins."

1. ***P. galacturus,*** (Cope) Jordan. MILKY-TAILED SHINER. SLENDER SILVER-FIN. Bluish above, sides bright silvery, with bright reflections; head 4⅓ in length; depth 4¾; mouth large, nearly horizontal; body slender, more elongated and less compressed than in the next; scales smooth and firm, usually with dusky edges; D. I, 8; A. I, 8; lat. l. 40; L. 5. Ohio Valley and S., abundant. Resembles the next, but larger, and with a larger mouth.

2. ***P. spirlingulus,*** (Val.) Jordan. SILVER-FIN. Leaden silvery; fins satin white in the breeding season; dorsal with a conspicuous black spot, as in the preceding; head 4 in length, rather short and deep; mouth rather small, very oblique, yet the lower jaw received within the upper in the closed mouth; body much compressed; depth 3¾ in length; D. I, 8; A. I, 8; lat. l. 35 to 40; L. 3½. Cayuga L., N. Y. (*S. H. Gage*), to N. J., Va., and Ind., abundant. In full breeding dress one of the most exquisite of all our fishes. (*Leuciscus spirlingulus*, Val. *L. kentuckiensis*, Kirt., not of Raf. *Rutilus plargyrus*, Raf. *Cyprinella analostana*, Grd.)

PHOTOGENIS, Cope. WHITE-CHEEKED SHINERS.

* Anal I, 10 or I, 9; no black caudal spot.

1. ***P. leucops,*** Cope. WHITE-CHEEKED SHINER. Dorsal fin much nearer caudal than end of snout; mouth very oblique; olivaceous above, sides silvery; scales large; eye white, 3 in head; head 4 to 4½ in length; depth less; D. I, 8; A. I, 10; lat l. 35 to 40. Ohio Valley; a handsome fish resembling a *Minnilus*.

2. ***P. arriommus,*** Cope. BIG-EYED SHINER. General appearance of preceding, but larger, reaching a length of nearly 5 inches; eye very large, 2½ in head, relatively larger than in any other of our Minnows; head large; bluish above, sides bright silvery; D. I, 8; A. I, 9; lat. l. 40. White R., Indiana, abundant, but not yet recognized elsewhere.

3. ***P. telescopus,*** Cope. WHITE SHINER. Dorsal fin equidistant; sea-green, silvery below; D. I, 8; A. I, 10; lat. l. 38. Holston R.

** Anal I, 8; a black spot at base of caudal.

4. ***P. leuciodus,*** Cope. WHITE SHINER. Olive, sides silvery with purple reflections; snout and base of dorsal red; lat. l. 40. Holston R.

*** Anal I, 8; no caudal spot.

5. ***P. scabriceps,*** Cope. ROUGH-HEADED SHINER. Head broad, minutely rough; eye large, 3 in head; head flattish above; mouth little oblique; greenish, sides leaden silvery; D. I, 8; A. I, 8; lat. l. 38. Ohio Valley.

6. ***P. spilopterus,*** Cope. SPOTTED-FINNED SHINER. Head narrower; eye smaller; olive, a plumbeous lateral band; black spot on dorsal behind; D. I, 8; A. I, 8; lat. l. 38. St. Joseph's R., Mich. This and some of the other species of this genus need confirmation.

20. MINNILUS, Rafinesque. ROSY-FACED MINNOWS.
= *Alburnellus*, Girard.

1. ***M. rubrifrons,*** (Cope) Jordan. ROSY-FACED MINNOW. Olive above, with a clear green lustre; sides silvery; a dark vertebral line; forehead, opercular region, base of dorsal, etc., flushed with red in spring; upper surface of head minutely tuberculate in males at that season; head rather pointed, about $3\frac{4}{5}$ in length; depth $4\frac{1}{2}$; eye about 4 in head; D. I, 8; A. I, 10; lat. l. 38; L. 3 or less. Ohio Valley, abundant; an elegant little fish, well distinguished from *M. rubellus* by the smaller size, deeper body and much longer head, as well as by peculiarities of form.

Var. ***amœnus,*** (Abbott) Jordan. ABBOTT'S ROSY MINNOW. Eye larger; head rather shorter. Delaware R.

2. ***M. dilectus,*** (Grd.) Jordan. DELECTABLE MINNOW. Intermediate between the preceding and the next; smaller than *rubellus* and more thick-set; head = depth,

about 5 in length; eye longer than snout, 3 in head; coloration of the others; D. I, 8; A. I, 11; lat. l. 42; L. 3½. Ohio R. (New Albany, Dr. *Sloan*) to Arkansas R. and S. (Type of *Alburnellus.*)

3. ***M. rubellus,*** (Ag.) Jordan. Rosy Minnow. Light olive, with brilliant clear green lustre; a dark vertebral line, and dark edges to the dorsal scales; sides brilliantly silvery, the lustre overlying a plumbeous lateral shade; forehead, etc., rosy in spring; sides sometimes rosy tinted; golden dorsal and lateral stripes, conspicuous in life as in most silvery species; head short, somewhat pointed, 5 in length; depth 5⅓ to 5½; eye 4 in head; D. I, 8; A. I, 10; lat. l. 38; L. 4 to 5. Great Lakes and Ohio Valley; abundant in the larger streams; even more graceful in form and delicate in coloration than the preceding.

4. ***M. dinemus,*** Raf. Emerald Minnow. Coloration exactly as in *M. rubellus*, but the body very slender and less compressed, more elongated than in any other of our *Cyprinidæ*, the depth being only from one-sixth to one-seventh of the length; head 4¾ in length; eye 3⅛ in head; fins as in preceding; L. 4 to 5. L. Michigan and Ohio Valley, in the larger streams, like the others, "going in flocks." (*A. jaculus* and *A. arge*, Cope.) (This is Rafinesque's "Emerald Minnow," the type of his genus *Minnilus.* Rafinesque's generic name having nearly forty years priority over *Alburnellus*, must be substituted for the latter appellation.)

5. ***M. micropteryx,*** (Cope) Jordan. Small-Finned Minnow. Resembles *M. rubrifrons*, but the fins all very low, the ventrals scarcely reaching to the line of the middle of dorsal; head 4½ in length; depth 5½ to 5¾; lat. L. 39; L. 3. Clinch R.

21. NOTEMIGONUS, Rafinesque. GOLDEN SHINERS.

= *Stilbe*, DeKay (preoccupied in Botany.)
= *Stilbius*, Gill (substitute for Stilbe.)
= *Luxilus*, Girard (not of Raf.)
= *Leucosomus*, Storer (not of Hæckel.)
= *Plargyrus*, Putnam (not of Raf.)
< *Abramis*, Cuvier (a closely related European genus).

1. ***N. americanus,*** (L.) Jordan. SHINER. STILBE. BREAM. Body much compressed; abdomen trenchant; head small, about 4 in length; depth 3 (2½ to 4); lateral line much decurved; scales small on the back, much larger below; dark steel blue or green above, sides silvery or golden, every where with brilliant reflections, green, yellow, and red; young specimens paler, looking like young *Luxili*, but the adults are among the largest in the family and bear a strong resemblance to Shad, a circumstance which has misled many observers, and among them Rafinesque; D. I, 7; A. I, 14; lat. l. 45 to 50. New England to Minnesota and S.; abundant in bayous, ponds, and weedy streams; this species is much more tenacious of life than any other of our *Cyprinoids*. [*N. auratus*, Raf. *S. chrysoleuca*, (Mitch.) DeK. *A. versicolor*, DeK.]

21. CARASSIUS, Nilsson. CRUCIAN CARPS.

1. ***C. auratus,*** (L.) Bleeker. GOLD FISH. Orange or blackish, rarely pale; D. I, 19; A. I, 8; lat. l. 26; exceedingly variable in domestication Asia; common every where in aquaria, and now naturalized in many of our eastern streams.

22. CYPRINUS, Linnæus. CARPS.

1. ***C. carpio,*** L. EUROPEAN CARP. Olivaceous;

D. III, 20; A. III, 5; lat. l. 37. European, introduced into some eastern rivers.

FAMILY CIX.—CATOSTOMIDÆ.

(*The Suckers.*)

Cyprinoid fishes of medium or large size, with the pharyngeal teeth in a single series, very numerous and closely set; intermaxillaries forming but a small part of the upper arch of the mouth, the maxillaries entering into it extensively on each side; mouth toothless, with fleshy lips, extremely protractile, roundish when fully protruded; dorsal fin long; anal short and rather high; no barbels; scales large; head naked; air bladder large, divided into two or three parts by transverse constrictions. Genera ten, *Pantosteus*, Cope, and the following; species about fifty, abounding everywhere east of the Rocky Mountains; two or three occur in China and Japan, all the rest are North American.

* Dorsal moderate, of 11 to 20 rays; size rarely large; Suckers. (CATOSTOMINÆ.)
  † Air bladder in two parts.
    ‡ Lateral line well developed; lips papillose.
      a. Scales much smaller anteriorly than posteriorly; interorbital space convex; body sub-terete. CATOSTOMUS, 1.
      aa. Scales about as large on front part of body as on tail; body tapering rapidly from shoulders to tail; interorbital space concave; length of head greater than depth of body; young barred or variegated. HYPENTELIUM, 2.
    ‡‡ No lateral line; lips usually plicate. . ERIMYZON, 3.
  †† Air bladder in three parts; lips usually plicate; lateral line very distinct.
    b. Pharyngeal teeth numerous and all small, of the usual type; the bones slender. . . . MOXOSTOMA, 4.

bb. Teeth on the lower half of the pharyngeal bones reduced to about a dozen, which are quite large; upper teeth numerous and small; bones stout and heavy. PLACOPHARYNX, 5.

** Dorsal elongated, of 20 to 35 rays, more or less elevated in front; size usually large; Buffalo fishes.

c. Body deep and heavy, usually considerably compressed; general color olivaceous white or brown. (BUBALICHTHYINÆ.)

d. Anterior dorsal rays very much elongated, reaching when depressed at least to the middle of the fin; back elevated; mouth small, inferior; pharyngeal teeth minute, nearly equal. . . . . CARPIODES, 6.

dd. Anterior dorsal rays moderately elevated, rarely reaching the middle of the fin; pharyngeal teeth becoming larger downward; eye small, operculum very large.

e. Mouth sub-terminal, protractile forwards. ICHTHYOBUS, 7.

ee. Mouth inferior, protractile downwards. BUBALICHTHYS, 8.

cc. Body very much elongated, not much compressed; dorsal of 35 rays; anal with 7 rays, placed very far back, the abdomen being therefore very long; general color black. (CYCLEPTINÆ.) . . . . . . CYCLEPTUS, 9.

1. ***CATOSTOMUS,*** LeSueur. BROOK SUCKERS

> *Acomus* and *Minomus*, Grd.

= *Decactylus*, Raf.

1. ***C. teres,*** (Mit.) LeS. COMMON SUCKER. WHITE SUCKER. Depth about equal to length of head, 4 to 4½ in length; olivaceous, sides silvery, with bright reflections; D. I, 13; A. 8 or 9; lat. l. 63. U. S., abundant everywhere E. of the Mississippi. (*C. communis*, *bostoniensis*, etc., of authors.)

2. ***C. hudsonius,*** LeS. NORTHERN SUCKER. LONG-NOSED SUCKER. Very slender; depth less than length

of head; snout very much produced; scales very small; D. I, 12; A. 9; lat. l. 105. L. Superior to Arctic regions.

2. *HYPENTELIUM,* Rafinesque. BIG STONE LUGGERS.

= *Hylomyzon*, Ag.

1. ***H. nigricans,*** (LeS.) Jordan. STONE ROLLER. MUD SUCKER. Depth $4\frac{3}{4}$ in length; head 4; depth of head $\frac{2}{3}$ its length; eyes small, very high up and far back; lower fins very large; pectoral nearly as long as head; brownish; often beautifully marbled; D. 12; A. 8; lat. l. 52. Lakes and streams from N. Y., S. and W., abundant; one of our most singular fishes. It frequents clear streams and rapids, and it is not at all a "mud fish." as some writers seem to suppose.

3. *ERIMYZON,* Jordan. CHUB SUCKERS.

= *Moxostoma*, Agassiz (not of Raf.)

* No stripes along the rows of scales.

1. ***E. oblongus,*** (Mit.) Jordan. CREEK FISH. CHUB SUCKER. Head 4 to $4\frac{1}{2}$ in length; depth $2\frac{3}{4}$, in adult; eye 5 in head; scales crowded, deeper than long; dusky above, brassy on sides and below; very variable; young much less compressed, with black bands or bars, and pale lateral and vertebral streaks; spring males with six tubercles on head; D. 12; A. 8; lat. l. 40. New England S. and W., abundant in Great Lakes.

** Each scale with a large, square black spot at its base, these forming conspicuous stripes along the sides.

2. ***E. melanops,*** (Raf.) Jordan. STRIPED SUCKER. Head $4\frac{2}{3}$ in length; depth about 4; scales very large; blackish above; sides coppery, with black stripes; D. I, 13; A. I, 8; lat. l. 47; size large. Great Lakes and

Ohio Valley, abundant; one of our handsomest suckers, strangely overlooked by recent writers. This and the preceding, unlike most of our suckers, are very hardy in the Aquarium.

3. ***E. sucetta,*** (Lac.) Jordan. LACEPEDE'S SUCKER. Head compressed and flat; lower lip very large; brown, sides silvery, with brown stripes along the rows of scales; D. I, 12; A. 9. Southern States.

4. ***MOXOSTOMA,*** Rafinesque. RED HORSES.

× *Teretulus*, Raf.

= *Ptychostomus*, Ag.

* Dorsal with 13 to 15 developed rays; body compressed.

† Lower fins reddish, becoming orange on death.

1. ***M. duquesnei,*** (LeS.) Jordan. COMMON RED HORSE. WHITE MULLET. Head 4 to 4⅔ in length; depth about 4½; eye large, 3½ to 4 in head; olive above, sides bright silvery, with red and green reflections; lower fins pink, becoming bright crimson; D. I, 13; A. 8; lat. l. 42 to 47. Ohio Valley and Lake region, every where abundant. [*Pt. erythrurus* (Raf.) Cope.]

2. ***M. aureolum,*** (LeS.) Jordan. LAKE RED HORSE. GOLDEN MULLET. Head quite small, about 5 in length; mouth large, not much inferior; eye 5 in head; yellowish brown, with bright reflections; lower fins decidedly red; back somewhat elevated; D. I, 13; lat. l. 49; large, reaches a weight of 20 lbs. Great Lakes, abundant.

3. ***M. anisurus,*** (Raf.) Jordan. CARP MULLET. Head stout, less than 4 in length; body short and thick, depth 3½; mouth with the lower lips decidedly **V**-shaped; eye small, 4¾ in head; lower fins pale orange; D. I, 13 to I, 15. North Carolina to Ind., and S. (*P. collapsus*, Cope.)

†† Lower fins white, never orange.

4. ***M. macrolepidotum,*** (LeS.) Jordan. LARGE-SCALED MULLET. Fusiform, compressed, depth 3½ in length; head short, convex, 4½ in length; eye large, 4 in head; color pale, with dusky and yellow shades; D. I, 13; V. 9; lat. l. 45. N. Y. to Ind., chiefly eastward.

** Dorsal with 16 to 18 developed rays; body compressed.

5. ***M. carpio,*** (Val.) Jordan. SILVERY MULLET. Head 4¼ in length; depth 3¼; eye 3⅓ in head; scales large, silvery white; lips large; dorsal larger than in any other species, dusky at tip; olivaceous, sides silvery, lower fins white; D. I, 8; V. 10; lat. l. 43. Great Lakes, not common. (Described from specimen from Fox R., Wis.)

6. ***M. velatum,*** (Cope) Jordan. Much like preceding, but lips as in *P. anisurus;* head short; D. 16; V. 9; lat. l. 42. Ohio Valley.

*** Dorsal with 12 rays; body nearly cylindrical.

7. ***M. cervinum,*** (Cope) Jordan. JUMPING MULLET. Head 5 in length; flattish above; lips large; yellowish with green reflections; size small; D. I, 12. Va., N. C., etc.; said to resemble *Hypentelium.*

5. PLACOPHARYNX, Cope. PLACOPHARYNX.

1. ***P. carinatus,*** Cope. COPE'S SUCKER. Resembles *M. duquesnei*, but the pharyngeal bones quite different; eye 4½ in head; head 4 in length; depth 3⅔; head strongly ridged above; pharyngeal bones very heavy, the lower 7 to 12 teeth on each side very large, truncate, irregularly placed; D. 14; A. 7; lat. l. 41. Wabash R.; probably not uncommon, but not distinguished until quite lately.

6. *CARPIODES*, Rafinesque. Carp Suckers.

* First rays of dorsal very much elevated and attenuated, about as long as the base of the fin.

1. ***C. velifer,*** (Raf.) Ag. Spear Fish. Sail Fish. Quillback. Skimback. Muzzle conic, much less obtuse than in the next; depth 2½ in length; head 3¾; eye 4¼ in head; color pale, scarcely silvery, as in all species; D. 22; lat. l. 37. Ohio R.

2. ***C. difformis,*** Cope. High-Backed Carp Sucker. Snout very blunt; eye 4½ in head; head 3¾ in length; depth 2½; body short and high; dorsal fin strongly falcate; snout minutely tuberculate in spring as in some other species; D. 26; A. 8; V. 9; lat. l. 37. Western streams (described from L. Erie specimens of "*C. cutisanserinus*," Cope = *C. selene*, Cope.)

** Anterior dorsal rays scarcely filamentous, little more than half the length of the base of the fin.

3. ***C. bison,*** Ag. Buffalo Carp. Muzzle very long, conic; eye large, 4½ in head; longest dorsal rays reaching nearly to end of fin; D. 28; lat. l. 40. Mississippi Valley and W.

4. ***C. thompsoni,*** Ag. Lake Carp. Short and stout; scales narrowly exposed; eye small, 5¼ in head; depth 2½ in length; long rays of dorsal reaching 22d ray; D. 28; lat. l. 41. Great Lakes.

5. ***C. cyprinus,*** (LeS.) Ag. Silvery Carp Sucker. Body oblong; dorsal as above; eye 5 in head; depth 3¾ in length; head 3; D. 30; lat. l. 40. Rivers, chiefly eastward.

6. ***C. carpio,*** (Raf.) Jordan. Olive Carp Sucker. Elongated; head small; dorsal rays short; depth 3 in length; eye 4⅗ in head; D. 30 or more; lat. l. 36; the largest species. Ohio Valley. (*C. nummifer*, Cope.)

7. ***ICHTHYOBUS,*** Rafinesque. Buffalo Fishes.

< *Sclerognathus*, Val.

1. ***I. bubalus,*** (Raf.) Ag. Brown Buffalo Fish. Depth $3\frac{1}{2}$ in length; head the same; eye small, $6\frac{1}{2}$ in head; depth of head five-sixths its length; opercle very wide, forming nearly half the length of head — convex and furrowed; scales very large; dull brownish olive, not silvery; D. 27; A. 10; lat. l. 40; length (of specimen) 27 inches; weight 15 lbs. Mississippi Valley (described from specimen from Wabash R.)

8. ***BUBALICHTHYS,*** Agassiz. Buffalo Suckers.

1. ***B. niger,*** (Raf.) Agassiz. Buffalo Fish. Depth = length of head, $3\frac{3}{4}$ in body; eye 4 to 6 in head; nape prominent, convex; length of opercle half its height; D. 30 to 35; A. 8. Mississippi Valley; known from the preceding by its more arched back, inferior mouth and probably greater size. Several other species of *Ichthyobus* and *Bubalichthys* have been mentioned but not recognizably described.

9. ***CYCLEPTUS,*** Rafinesque. Suckerels.

= *Rhytidostomus*, Hæckel.

1. ***C. elongatus,*** (LeSueur) Ag. Black Horse. Gourd-Seed Sucker. Long Buffalo. Missouri Sucker. Body fusiform, not greatly compressed; mouth very small; depth 4 to 5 in length; lobes of dorsal and caudal much attenuated; jet black above, sides black with a coppery lustre; snout minutely tuberculate in spring; D. 35; length 2 to 3 feet; weight 2 to 15 lbs. Mississippi Valley, in large streams. A singular species, quite unlike any other.

ORDER AA.—NEMATOGNATHI.

(The Sheat Fishes.)

Skin naked or with bony plates; no true scales; barbels always present, maxillary bone rudimentary and forming the base of the longest barbel; margin of upper jaw formed by intermaxillaries only; sub-opercle absent; air bladder generally present; usually an adipose fin, and in all our species a spine in the dorsal and pectorals; ventrals abdominal. Chiefly fresh water fishes, apparently related to the Sturgeons. The leading family is *Siluridæ*.

FAMILY CX.—SILURIDÆ.

(The Cat Fishes.)

General characters as above given,—ours all have the air bladder well developed, the skin naked, and the dorsal and pectoral spines more or less developed and often serrated behind; the fresh water species have always eight barbels. Genera one hundred or more; species nearly seven hundred; a very large family abounding in the fresh waters of America and the warmer parts of the Old World; a few are marine. Our species, though very numerous, are closely related.

* Eyes well developed.

† Adipose fin free from caudal.

‡ Dorsal spine well developed; branchiostegals less than 12.

a. Supra-occipital bone connected behind with the base of the dorsal spine; body slender, elongated; head relatively small; tail strongly forked. . ICTALURUS, 1.

aa. Supra-occipital bone free behind; body short and thick; head very large, depressed; caudal rounded (forked in two or three species.) . . . AMIURUS, 2.

‡‡ Dorsal spine rudimentary; branchiostegals 12; body elongated; lower jaw longest. . . HOPLADELUS, 3.

†† Adipose fin long, keel-like, continuous with the rounded caudal; spines well developed. . . NOTURUS, 4.

** Eyes concealed beneath the skin; blind cave fishes. GRONIAS, 5.

1. ***ICTALURUS,*** Rafinesque. CHANNEL CATS.

= *Synechoglanis*, Gill.

1. ***I. punctatus,*** (Raf.) Jordan. CHANNEL CAT. BLUE CAT. WHITE CAT. Depth 5 in length; head 4¼, longer than broad; upper jaw longest; clear olivaceous, silvery below and on sides; young punctate with black; barbel longer than head; D. I, 6; A. 26; P. I, 8. Mississippi Valley, abundant. [*I. cærulescens*, (Raf.) Gill.]

2. ***I. nigricans,*** (LeS.) Jordan. GREAT LAKE CAT FISH. Olive brown, sides ashy with large faint black blotches; back nearly black; D. I, 6; P. I, 9; A. 26; length 2 to 4 feet. Great Lakes, abundant; reaches a large size. (This species proves to be an *Amiurus;* see Addenda.)

3. ***I. furcatus,*** (LeS.) Gill. GREAT FORK-TAILED CAT. Olivaceous, silvery; back blackish; maxillary barbel shorter than head; D. I, 7; P. I, 10; A. 33; very large. Mississippi Valley.

4. ***I. furcifer,*** (C. & V.) Gill. FORK-TAILED CAT. Olivaceous; barbels longer than head; D. I, 6; P. I, 9; A. 27. Mississippi Valley; a doubtful species.

5. ***I. gracilis,*** (Hough) Gill. NORTHERN SILVERY CAT. D. I, 5; A. 23; barbels as long as head. N. N. Y.

2. ***AMIURUS,*** Rafinesque. HORNED POUTS.

< *Pimelodus*, early authors.

* Caudal fin deeply forked.

1. ***A. lynx,*** (Girard) Gill. POTOMAC CAT. Width of head 4½ in length; eye large, 3 to 5 in interorbital space;

barbels long; black above, sides silvery, belly pure white; D. I, 6; A. 22; V. 8. Rivers of Atlantic Coast, N. J. to S. C.

2. ***A. lophius,*** Cope. BIG-MOUTHED CAT. Width of head 3½ in length of body; head and mouth excessively large; brown, whitish below; D. I, 6; A. 21; V. 8. Potomac R., etc.

** Caudal fin deeply emarginate, but not forked.

3. ***A. confinis,*** (Girard) Gill. WISCONSIN BULL-HEAD. Dorsal nearer adipose fin than snout; jaws equal; D. I, 6; A. 20. Root R., Wis.

4. ***A. hoyi,*** (Grd.) Gill. HOY'S BULL-HEAD. Dorsal nearer snout than adipose fin; upper jaw longest; D. I, 7; A. 23. Racine, Wis. (Same as preceding?)

*** Caudal rounded or truncate when spread open (rarely very slightly emarginate.)

† Anal with 17 rays; upper jaw longest.

5. ***A. pullus,*** (DeK.) Gill. BLACK BULL-HEAD OF N. Y. Blackish above; pale beneath; size small; head deep, 3½ in length; depth same; D. I, 5; P. I, 7; A. 17. Lakes of New York and eastward, abundant.

†† Anal with 19 to 22 rays.

‡ Jaws about even, or upper jaw projecting.

6. ***A. atrarius,*** (DeK.) Gill. NORTHERN BULL-HEAD. HORNED POUT. Head 4 in length; depth 4⅓; width of head 5½ to 6; barbels long; slope nearly uniform upward from snout to the elevated base of dorsal, a character not shown by any other species known to me; blackish above, sides coppery, belly white, yellowish forwards; size small; D. I, 6; P. I, 8; A. 20. New England to Maryland and the Great Lakes, abundant; the common eastern species.

7. ***A. albidus,*** (LeS.) Gill. BROWN CAT FISH. WESTERN BULL-HEAD. Depth 3½ to 4¾ in length; head 3¾ to 4½, its width 5; barbels moderate; eye 7 in head; head wider than in the preceding; an uneven curve from snout to base of dorsal which is not especially elevated; yellowish brown or blackish above, sides coppery yellow, belly usually decidedly yellow but variable, rarely blackish and pale; caudal truncate; D. I, 7; P. I, 8; A. 19 to 22. Lakes and streams, abundant; N. Y. to Minn., Kansas, and S.; chiefly west of the Alleganies. [*A. nebulosus* (LeS.), *A. catulus* (Grd.), *A. xanthocephalus* (Raf.), etc.] Perhaps two or more species are here confounded.

‡‡ Lower jaw longer than upper.

8. ***A. dekayi,*** (Girard) Gill. DEKAY'S BULL-HEAD. Form nearly of *A. atrarius;* head and body rather elongated; A. 20, its base one-sixth of length of fish. Lakes, etc., in N. Y. A doubtful species. ["*P. catus,*" DeK., *fide* Girard. *A. catus* (L.), Gill, is a Southern species.]

9. ***A. cœlurus,*** (Girard) Gill. MINNESOTA CAT FISH. Head broad, 3½ in length; body rather short; reddish, pale below; D. I, 6; A. 22; its base more than ⅕ of length. L. Amelia, Minn. to Ills. (*Forbes*) and S.

††† Anal with 24 to 28 rays; size usually large.

10. ***A. cupreus,*** (Raf.) Gill. GREAT YELLOW CAT FISH. Upper jaw longest; head very large; body stout; barbel rather short, but usually as long as head; color usually a clear copper yellow, belly yellow; reaches a very large size, probably the largest in the genus; D. I, 6; P. I, 7; A. 25. Lakes and larger Western streams, abundant.

11. ***A. lividus,*** (Raf.) Jordan. LIVID CAT FISH. Jaws

equal; color dark; size small; otherwise like the preceding. Ohio to Ills. (needs confirmation.) [*A. cupreoides*, (Grd.) Gill.]

12. ***A. cœnosus,*** (Rich.) Gill. LAKE HURON CAT FISH. Dark greenish, paler below; spines strongly serrated; D. I, 7; P. I, 8; A, 24. Lake Huron. A doubtful species.

3. *HOPLADELUS,* Rafinesque. MUD CATS.

1. ***H. olivaris,*** (Raf.) Gill. MUD CAT FISH. Body elongated, long and low; lower jaw longest; head depressed, thick-skinned; caudal emarginate; olive brown, paler below; size very large; D. I, 6; P. I, 10; A. 15 (or more?). Mississippi Valley. (Other species probably exist, but they have never been studied.)

4. *NOTURUS,* Rafinesque. STONE CATS.

1. ***N. flavus,*** Raf. OHIO STONE CAT. Head 3 to 4 in length; dorsal longer than high; barbels rather short; color nearly uniform yellowish brown; body thick and short; D. I, 7; P. I, 7; A. 16. Ohio Valley; abundant and very variable.

2. ***N. lemniscatus,*** (LeS.) Gill. SOUTHERN STONE CAT. Head broad, 4¼ in length; spines short; D. I, 7; P. I, 10; A. 21. Rivers S. & W.

3. ***N. marginatus,*** Baird. MARGINED STONE CAT. Head 4 in length; dorsal higher than long; barbels rather long; brownish or blackish, much mottled and barred; fins black-edged; head smaller than in *N. flavus*, and the body more elongate; D. I, 7; A. 14; P. I, 7. Penn. to S. C. and W.

4. ***N. gyrinus,*** (Mit.) Raf. NORTHERN STONE CAT. Head elongate, flat, 3¼ in length; barbels long; brown-

ish, fins yellow-edged; D. I, 6; P. I, 6; A. 16. Walkill R., N. Y., etc. (The species of this genus need a critical revision.)

5. *GRONIAS*, Cope. Blind Cats.

1. ***G. nigrilabris***, Cope. Blind Cat Fish. Form of *Amiurus;* blackish; D. I, 7; P. I, 9. Subterranean stream tributary to Conestoga R., E. Penn.

ORDER BB. — APODES.

(*The Eels.*)

Maxmary bones rudimentary; body serpentiform; no ventral fins; scapular arch free from skull; scales small or wanting.

FAMILY CXI. — ANGUILLIDÆ.

(*The Eels.*)

Body much elongated, nearly cylindrical, covered with small scales; vent posterior; pectorals present; vertical fins confluent; sides of upper jaw formed by the maxillaries; intermaxillaries more or less coalescent with the vomer; stomach cœcal. Genera two or three; "an infinite number of species have been described" (*Günther*), but the actual number cannot exceed forty. In seas and sluggish fresh waters of all regions.

1. *ANGUILLA*, Thunberg. Eels.

1. ***A. rostrata***, (LeS.) DeK. Common American Eel. Distance between dorsal and anal shorter than head. U. S., chiefly coastwise, but ascending all rivers and introduced into the Great Lakes. But one species is recognizable. (*A. bostoniensis*, Authors, but the name *rostrata* has priority.

Sub-Class Ganoidei.

The Ganoid Fishes.

Skeleton bony or cartilaginous; tail more or less heterocercal; optic nerves forming a chiasma; arterial bulb rhythmically contractile, provided with several rows of valves; air bladder frequently cellular and lung-like; skin usually with bony plates; intestine usually with a spiral valve; ventral fins, if present, abdominal. Of this important sub-class but few species are now existing, and these few vary widely from one another. Of the earlier fossil fishes, a very large proportion are Ganoids.

ORDER CC.—CYCLOGANOIDEI.

(*The Cycloganoids.*)

This order contains but a single species among recent fishes.

FAMILY CXII.—AMIIDÆ.

(*The Bow-Fins.*)

Body oblong, rather stout, covered with thick cycloid scales; tail heterocercal, the caudal peduncle curved upwards behind, "like a sled-runner;" a large bony buckler between branches of lower jaw; membrane bones of head much developed, very hard; jaws broad with strong teeth in two sets, similar teeth on vomer, palate and pterygoids; snout short, rounded; ventrals large, abdominal; dorsal very long, the rays of nearly uniform height; anal short and rather high; air bladder large, cellular, lung-like, communicating by a glottis with the œsophagus; stomach large.

A single species is found in the larger bodies of fresh water in the U. S. from N. Y. to the plains. It is exceedingly tenacious of life, even more so than the species of *Amiurus*. The flesh is soft and pasty, and not edible. In some regions its voracity has acquired for it the name of "Lawyer," because, as has been said, "it will bite at any thing, and is good for nothing when caught."

1. AMIA, Linnæus. Bowfins.

1. ***A. calva,*** L. Bowfin. Dog Fish. Mud Fish. Depth 4 to 4½ in length; head nearly 4; eye 8 in head; anterior nostrils each with a short barbel; dark olive or blackish above, nearly white below; sides with traces of greenish markings; lower jaw and gular plate with round blackish spots; fins mostly dark; ♂ reaching a length of 18 inches, with a roundish black spot on the upper base of caudal, which is surrounded by an orange or yellowish shade; ♀ larger, 2 feet or more in length, without the black caudal spot; D. 42 to 53; A. 10 to 13; lat. l. 65 to 70. E. U. S.; abundant in the Great Lakes. (*A. ornata*, *reticulata*, and *viridis*, LeS. *A. marmorata*, *canina*, *lentiginosa*, *cinerea*, and *subcærulea*, Val. *A. ocellicauda*, Rich. *A. occidentalis*, DeK. *A. thompsoni* and *piquotii*, Duméril.)

ORDER DD.—RHOMBOGANOIDEI.

(*The Rhomboganoids.*)

This order includes, among recent fishes, only the following family:

FAMILY CXIII.—LEPIDOSTEIDÆ.

(*The Gar Pikes.*)

Body elongated, sub-cylindrical, covered with hard, enamelled, lozenge-shaped, ganoid plates; snout elongated, spatulate, or beak-like; upper jaw of several pieces, longer than the lower, which is formed of as many parts as in Reptiles; both jaws and palate armed with bands of rasp-like teeth, and series of larger, conical ones; fins with fulcra (elongated modified scales) in front; dorsal and anal short and placed far back, moderately high; vertebræ concavo-convex, with ball and socket joints as in Reptiles; air bladder cellular, like the lungs of Reptiles, connected with the pharynx; stomach not cœcal but with numerous pyloric appendages; intestine with rudimentary spiral valve; no spiracles; branchiostegals three; pseudobranchiæ present. Fresh waters of N. A., from New England to the Rocky Mountains, S. to Central America and Cuba. Genera two or three (*Cylindrosteus* seems to us to be rather a sub-genus of *Lepidosteus*); species probably about five although forty have been described; until some more tangible distinctions are shown, we can admit but one in each genus.*

* In a recent work on these fishes, Prof. August Dumeril very laboriously distinguishes the following "species" among the specimens of *Lepidosteus* in the Museum at Paris:

L. osseus, (L.) (= *L. gavialis*, Lac.); *L. louisianensis*, Dum. (= *L. oxyurus*, Raf. = *Sarchirus vititatus*, Raf.); *L. harlani*, *L. ayresi*, *L. smithii*, *L. copei*, *L. lamarii*, *L. clintonii*, *L. troostii*, *L. piquotianus*, *L. lesueurii*, *L. elizabeth*, *L. thompsoni*, *L. horatii*, *L. milberti*, *L. treculii*, Dumeril; and *L. huronensis*, Rich. Of *Cylindrosteus*, he finds *C. platystomus* (Raf.); *C. productus* (Cope); *C. platyrhynchus* (DeK.); *C. agassizii*, *C. rafinesquei*, *C. bartoni*, *C. castelnaudii* and *C. zadocki*, Dum.

Most of these nominal species are based upon the most trifling individual differences, and often the right side of a specimen indicates one "species," and the left another. As matters stand, we have no alternative but to reject them all, and to wait for the time when systematic writers shall be wiser or more honest.

* Large teeth on the maxillaries in a single row; species of moderate size, 2 to 5 feet long.

† Snout very slender, straight, much longer than the rest of the head. . . . . LEPIDOSTEUS, 1.

†† Snout shortened, rather broad, about as long as rest of head. CYLINDROSTEUS, 2

** Large teeth on the maxillaries in two rows; snout broad, depressed, about equal to rest of head; size large, length 5 to 10 feet. . . . . LITHOLEPIS, 3.

1. *LEPIDOSTEUS,* Lacepede. LONG-NOSED GARS.

> *Sarchirus*, Raf. (Young with the pectoral fins fleshy.)

1. ***L. osseus,*** (L.) Ag. GAR PIKE. BONY GAR. BILL FISH. Head nearly 3 in length; depth nearly 12; snout more than twice the length of rest of head; eye nearly 2½ in distance to margin of preopercle, more than 2 in interorbital space; ventrals midway between pectorals and anal; length of mandible equals distance from pectorals to anal; olivaceous, white below; sides with obscure spots, more evident posteriorly; vertical fins with distinct round black spots; D. 7; A. 9; lat. line 64 to 66; length 2 to 5 feet. N. Y. to the plains and South, abundant in large bodies of water.

2. *CYLINDROSTEUS,* Rafinesque. SHORT-NOSED GARS.

< *Lepidosteus*, Agassiz.

1. ***C. platystomus,*** (Raf.) Grd. SHORT-NOSED GAR PIKE. Depth 7 to 8 in length; head 3½; eye 10 in head, three times nearer opercular margin than end of snout; ventrals much nearer P. than A.; length of head notably shorter than from P. to V.; olivaceous, sides and fins spotted with black; D. 7; A. 8; lat. l. 60 to 65. Great Lakes and streams S. and W. of N. Y. to Rocky Mountains.

3. *LITHOLEPIS,* Rafinesque. Alligator Gars.

= *Atractosteus*, Raf.

1. ***L. adamantinus,*** Raf. Great Gar. Alligator Gar. Diamond Fish. Snout broad, depressed, the length of the cleft of the mouth being about half the length of the head; color olivaceous; very large, reaching a length of of 8 feet or more. Mississippi Valley, N. to Illinois and Ohio, abundant southward. (*A. ferox*, Raf., but the name here adopted has precedence. Perhaps some still older specific name may be found.)

ORDER EE.—SELACHOSTOMI.

(*The Spoon-Bills.*)

This order contains but the single family *Polyodontidæ.*

FAMILY CXIV.—POLYODONTIDÆ.

(*The Spoon-Billed Cats.*)

Body elongated; skin naked, with minute stellated roughnesses, and some bony plates about head and tail; mouth very wide, not inferior but overhung by the long snout; minute teeth on lower jaw, maxillaries and palate, teeth sometimes deciduous with age; snout produced into a very long and spatula-like process, thin and flexible at its edges; no barbels; caudal with fulcra, as in *Lepidosteus*, heterocercal, the lower lobe well developed; opercle with a long flap reaching to pectorals, or beyond and sometimes to ventrals; spiracles present; no tongue; one broad branchiostegal; air bladder large, communicating with the œsophagus; intestine with a well developed spiral valve; stomach cœcal, with a broad divided pyloric appendage. Fresh waters of U. S. and China. Species two; *P. folium* from the Mississippi, and *P. gladius* from the Yangtsekiang.

1. POLYODON, Lacepede. Duck-Billed Cats.

= *Spatularia*, Shaw.
= *Planirostra*, LeSueur.

1. ***P. folium,*** Lacepede. Duck-Billed Cat. Spoon-Billed Sturgeon. Snout nearly $\frac{1}{3}$ of length; opercular flap reaching much beyond pectorals; fins all more or less falcate; color gray; D. 55 to 60; A. 56; length 5 feet or more. Mississippi and its larger tributaries, abundant. A singular fish, bearing considerable resemblance to a Shark.

ORDER FF.—CHONDORSTEI.

(*The Cartilaginous Ganoids.*)

This order is equivalent to the family *Acipenseridæ.*

FAMILY CXV.—ACIPENSERIDÆ.

(*The Sturgeons.*)

Body elongate, sub-cylindrical, with five longitudinal rows of bony shields, the lower sometimes deciduous; snout produced; mouth entirely inferior, transverse, protractile, toothless; four barbels in a row under snout in front of mouth; vertical fins with fulcra; caudal heterocercal; dorsal and anal far back. No branchiostegals; air bladder large, not cellular; stomach not cœcal, with pyloric appendages; intestines with spiral valve; distribution same as that of *Salmo.* Fresh waters of northern regions, some species marine and entering rivers. Genera two; species twenty (*Günther*), eighty or more

(*August Duméril*). Perhaps in no group has the making of nominal species been carried to a greater extent than in this.

* Rows of bony bucklers distinct throughout; spiracles present; snout rather narrow, sub-conical. . . ACIPENSER, 1.

** Rows of bony bucklers confluent behind, entirely surrounding the depressed tail; no spiracles; snout flattened, rather broad, shovel-shaped. . . . . SCAPHIRHYNCHUS, 2.

1. ACIPENSER, Linnæus. STURGEONS.

* Marine species ascending rivers; lateral shields 22 to 32.

1. ***A. sturio***, L. COMMON SEA STURGEON. SHARP-NOSED STURGEON. Snout pointed, half the length of head; dorsal shields 11 to 13; lateral shields 26 to 31; D. 37 to 44 rays. Atlantic Ocean S. to Africa and West Indies. (*A. oxyrhynchus*, Mit., the American form.)

2. ***A. brevirostris***, LeSueur. BLUNT-NOSED STURGEON. Snout blunt, one-quarter length of head; dorsal shields 8 to 10; lateral shields 22 to 28; abdominal 8 to 10; D. 30 rays; skin with minute scattered prickles and stellate ossifications. Cape Cod to Fla.

** Species of fresh water; lateral shields 33 to 38.

3. ***A. maculosus***, LeSueur. ROCK STURGEON. BONY STURGEON. Snout pointed, nearly as long as rest of head; head 3⅓ in length of body; bony shields large, close together, 13 to 16 in front of dorsal, 33 to 38 on sides, 9 or 10 on abdomen, all of them rough and strongly radiated, with more or less hooked or incurved tips; skin rough; D. 37 to 45. Great Lakes, Ohio R. and northward. (Possibly the young of the next ?)

4. ***A. rubicundus***, LeSueur. RED STURGEON. LAKE STURGEON. Head 4½ in length; eye 10 in head, nearly midway; dorsal scutes 16 (to base of D), relatively small

and rather distant; lateral scutes 35; ventral scutes 9; snout rather blunt, becoming more so with age, rather shorter than rest of head; barbels nearer to end of snout than to eye; scutes relatively smaller, smoother and less crowded than in the preceding; blackish, sides paler or reddish; length 2 to 6 feet; D. 37. Great Lakes and Western Rivers.

2. SCAPHIRHYNCHUS, Hæckel. SHOVEL-NOSED STURGEONS.

1. ***S. platyrhynchus,*** (Raf.) Grd. SHOVEL-NOSED STURGEON. Tail wider than deep, extending beyond caudal rays and ending in a filament; snout nearly the form of a spade; whole body rough with small prickles; dorsal shields 15 or 16; lateral shields 40 to 46; abdominal 10 to 13; all the shields rough and strongly carinated, the keel ending behind in a spine which points backward; size large. Mississippi Valley.

Class VII.—Marsipobranchii.

(*The Lampreys.*)

Skeleton cartilaginous, without ribs, limbs, shoulder girdle nor pelvic elements; skull imperfectly developed, without true jaws; a single median nostril; gills in the form of fixed sacs, without branchial arches; gill sacs typically seven on each side; mouth nearly circular, suctorial; no scales; body elongated, eel-shaped; alimentary canal nearly straight and simple; no arterial bulb.

(Class VI.—ELASMOBRANCHII, the *Selachians*, represented on our Atlantic Coast by many species of Sharks and Skates, is here omitted, as its members are exclusively marine, and it does not therefore come within the scope of this treatise. Class VIII.—LEPTOCARDII, the *Lancelets*, is also omitted for the same reason. The latter class contains, as far as now known, but a single genus, BRANCHIOSTOMA or *Amphioxus*, with two or three species. One of these, *Branchiostoma caribæum*, Sundevall, occurs along the coast of our South Atlantic States.)

ORDERS OF MARSIPOBRANCHII.

* Nasal duct a blind sac, not penetrating the palate.
HYPEROARTIA, GG.

** Nasal duct penetrating the palate. HYPEROTRETA, page 313.

GG. FAMILIES OF HYPEROARTIA.

* Branchial sacs seven on each side· intestine with spiral valve.
PETROMYZONTIDÆ, 116.

FAMILIES OF HYPEROTRETA.

* One external aperture on each side of body, leading by six ducts to as many branchial sacs; no spiral valve; marine parasites, burrowing into the bodies of other fishes.
Myxinidæ, the *Hag Fishes*.

ORDER GG.—HYPEROARTIA.

(*The Lampreys.*)

FAMILY CXVI.—PETROMYZONTIDÆ.

(*The Lampreys.*)

Body eel-shaped, naked; dorsal and anal fins long and low, usually continuous with the caudal; mouth suctorial, armed with horny teeth which rest on papillæ. Eggs small.

These animals undergo a metamorphosis; the young are usually toothless, and have the eyes rudimentary. Genera five or six, in temperate regions, found in all waters. They attach themselves to fishes and other animals, and feed by scraping off the flesh, by means of their rasp-like teeth.

The American species are still very imperfectly known. Until quite recently the *larvæ* were considered as forming distinct genera, which have been termed *Ammocœtes*, *Scolecosoma*, *Chilopterus*, etc.

* Dorsal fins two, nearly or quite distinct, the second continuous with the caudal; "maxillary tooth bicuspid."
PETROMYZON, 1.

** Dorsal fin single, continuous with the caudal; "maxillary tooth tricuspid." . . . . . ICHTHYOMYZON, 2.

1. ***PETROMYZON,*** Linnæus. LAMPREYS.

> *Ammocœtes*, Duméril, (*Larvæ.*)

1. ***P. marinus,*** (L.) var. ***americanus,*** (LeS.) Wilder. GREAT SEA LAMPREY. Resembles the next, but larger, with a shorter head, which is but little longer than the "chest" (space occupied by the branchial openings); color olive brown, mottled with black; L. 30 to 40. Marine, ascending rivers, eastward.

2. ***P. nigricans,*** LeS. LARGE BLACK LAMPREY. Head very large, longer than the "chest," 6½ in length; depth about 13; body little compressed; dorsal fins rather low, distinctly separated; eyes and mouth very large; a depression between eyes and snout; a single bicuspid tooth in front of œsophagus; mandibulary plate curved, with about eight pointed teeth; rest of mouth covered with rather large teeth disposed in oblique cross-rows, five or more in each row; lips fringed; L. 12. Lakes and streams, Cayuga L., N. Y. (*Wilder*), and E.; properly a marine species. (Several other Lampreys have been described from our Eastern streams and coast, but they are very doubtful.)

3. ***P. niger,*** Raf. SMALL BLACK LAMPREY. Head moderate, longer than "chest," 8⅓ in total length; depth 14; body scarcely compressed except behind; dorsal fins rather high, slightly connected; eyes large; mouth rather small, two rather large teeth well apart in front of œsophagus; mandibulary plate nearly straight, with about eight sub-equal teeth; a few scattering teeth on sides of mouth; snout rounded; dark blue black, unspotted, silvery below; L. 8 to 11. Great Lakes, Upper Miss. and Ohio Valley, abundant in many localities, ascending small brooks in the spring.

2. *ICHTHYOMYZON,* Girard. LAMPERNS.

> *Scolecosoma*, Grd. (*Larvæ.*)

1. ***I. argenteus,*** (Kirt.) Grd. SILVERY LAMPREY. Head quite small, shorter than "chest," about 10 in length; depth nearly the same, in adult; body stout, compressed; dorsal fin very high, with a shallow depression; eyes distinct in adult, concealed in young; mouth small, with inconspicuous teeth; color ashy silvery, with numerous

small black dots; larger ones above the gill openings; L. 12. Great Lakes and Ohio Valley, E. to N. Y., frequent. (*Ammocœtes concolor*, Kirt., *A. œpyptera*, Abbott.)

2. ***I. castaneus,*** Grd. Chestnut-colored without spots; eyes inconspicuous; "mandibulary plate with nine teeth." Minn.

3. ***I. hirudo,*** Grd. Grayish, unspotted; eyes small; "mandibulary plate with seven teeth." Ark.

ADDENDA.

(Page 233.)

Glossoplites melanops, (Grd.) Jordan. Two fine specimens, 6 inches long, of this species, taken in Lake Michigan, June, 1876, show the following characters in addition to those given in the text:

Dark olive green above; sides greenish and brassy with blotches of pale blue and bright coppery red—the latter shade predominating; belly bright brassy yellow, profusely mottled with bright red; lower jaw chiefly yellow; iris bright red, as in most Sun Fishes; opercular spot as large as eye, black, bordered with copper color; three or four wide dark red bands radiating backwards from eye across cheeks and opercles, separated by narrow pale blue ones; upper fins barred with black, orange and blue, the former color predominating; lower fins blackish; a small faint black spot on last rays of dorsal; dorsal spines moderately high, very stout, the longest as long as from snout to middle of eye; head $2\frac{2}{3}$ in length; depth $2\frac{1}{3}$; eye as long as snout, 4 to 5 in head; mucous pores about head very large.

(Page 235.)

Ichthelis incisor, (C. & V.) Holbr. Some very large specimens, taken in Lake Michigan, have the depth considerably more than half length, the flap very large and broad, sometimes an inch square, and the dorsal spot quite obscure. Occasionally the body is almost orbicular in form, the profile being nearly vertical. Some specimens have the belly strongly tinged with orange red.

(Page 300.)

"Ictalurus nigricans." The great Fork-Tailed Cat-Fish of the Lakes, mentioned in the text under the above name, is an *Amiurus*, and should stand as ***Amiurus nigricans,*** (LeS.) Gill.

A specimen two feet long, taken in Lake Michigan, shows the following characters: Tail strongly forked; supra-occipital bone free behind; head small, 4¼ in length, its width 5; upper jaw much longer than lower; barbels very long, longest reaching much beyond opercle; body heavy and deep, the depth being about 4 in length; body wider than head; dorsal outline elevated; color blackish, somewhat mottled, white below; D. I, 6; P. I, 10; A. 25. (*A. nigricans* of Günther and of Cope is described as having the caudal fin truncate which it certainly is not in this species.)

GLOSSARY

OF THE

PRINCIPAL TECHNICAL TERMS

USED IN THIS WORK.*

Abdomen—Belly.

Abdominal—Pertaining to the belly—said of the ventral fins of fishes when inserted considerably behind the pectorals, away from the shoulder girdle.

Abortive—Remaining or becoming imperfect.

Acuminate—Tapering gradually to a point.

Acute—Sharp-pointed.

Adipose fin—A peculiar, fleshy, fin-like projection on the backs of Salmons, Cat Fishes, etc., behind the dorsal fin.

Air bladder—A sac filled with air, lying beneath the back-bone of fishes, corresponding to the lungs of the higher vertebrates.

Allantois—An organ of the embryo.

Altrices—Birds reared in the nest and fed by the parents.

Altricial—Having the nature of *Altrices*.

Alula—The feathers attached to the "thumb" of a bird.

Alveolar surface—A portion of the jaw of a turtle, where the teeth-sockets (alveola) might be expected to be.

Amnion—An organ of the embryo.

Amphicœlian—Biconcave—said of vertebræ.

Anadromous—Said of marine fishes which run up rivers to spawn.

Anal—Pertaining to the anus or vent.

Anal fin—The fin on the median line, behind the vent, in fishes.

Anal plate—The plate, immediately in front of the vent, in serpents.

* In the preparation of this Glossary, the author has largely drawn from Dr. Coues' "Glossary of the Technical Terms used in Descriptive Ornithology," in Baird, Brewer and Ridgway's History of North American Birds, Vol. III. pp. 535—560.

Anteorbital plate—The plate, in front of the eye, in serpents.

Antrorse—Turned forwards.

Anus—The external opening of the intestines.

Arterial bulb—The muscular swelling, at the base of the great artery, in fishes, etc.

Articulated—Jointed.

Artiodactylous—Even-toed (toes 2 or 4.)

Attenuate—Long and slender, as if drawn out.

Auricle—The large lobe of the external ear; also, one of the chambers of the heart.

Barbel—An elongated fleshy projection, usually about the head, in fishes.

Basal—Pertaining to the base; at or near the base.

Beak—The bill of birds, or (in other animals) any beak-like structure.

Bend of Wing—Angle at the carpus when the wing is folded.

Bicolor—Two-colored.

Booted—Said of the tarsus, when its scales coalesce and form a continuous envelope.

Branchiæ—Gills; respiratory organs of fishes, etc.

Branchial—Pertaining to the gills.

Branchiostegals—The bony rays supporting the branchiostegal membranes, under the head of a fish, below the opercular bones, and behind the lower jaw.

Bristle—A stiff hair, or hair-like feather.

Caducous—Falling off early.

Calcareous—Containing or composed of carbonate of lime.

Canines—The teeth behind the incisors — the "eye-teeth"; in fishes, teeth in the front part of the jaws, longer than the others.

Carapace—The upper part of the shell of a turtle.

Carinate—Keeled, having a ridge along the middle line.

Carpus—The wrist.

Caudal—Pertaining to the tail.

Caudal fin—The fin on the tail of fishes.

Caudal peduncle—The region between the anal and caudal fins in fishes.

Cavernous—Containing cavities, either empty or filled with a mucous secretion.

Cere—Fleshy, cutaneous or membranous covering of the base of the bill in many birds, particularly the Owls, Hawks, and Parrots.
Cervical—Pertaining to the neck.
Chiasma—Crossing of the fibres of the optic nerve.
Chin—The space between the forks of the lower jaw.
Ciliated—Fringed with eye-lash-like projections.
Cinereous—Ashy in color.
Clamatorial—Pertaining to or like the *Clamatores*.
Clavicle—The collar bone.
Cœcal—Of the form of a blind sac.
Cœcum—An appendage of the form of a blind sac, connected with the alimentary canal.
Commissure—The line on which the mandibles of a bird are closed.
Compressed—Flattened laterally.
Condyle—Articulating surface of a bone.
Conirostral—Said of a bill like that of a Sparrow; conical in form and with the commissure angulated.
Costal folds—Folds of the skin (of a Salamander) showing the position of the ribs.
Crest—In birds, any lengthened feathers about the head; elsewhere, any elevated or crest-like projection.
Crissum—The under tail coverts, in birds.
Ctenoid—Rough-edged, said of scales when the posterior margin is minutely spinous or pectinated.
Culmen—The middle line or ridge of the upper mandible in birds.
Cuneate—Wedge-shaped; said of a bird's tail when the middle feathers are longest and the rest regularly shorter.
Cycloid—Smooth-edged; said of scales not ctenoid, but concentrically striate.
Deciduous—Temporary, falling off.
Decurved—Curved downward.
Dentate—With tooth-like notches.
Dentirostral—Having the bill notched near its tip.
Depressed—Flattened vertically.
Depth—Vertical diameter (usually of the body of fishes.)
Dermal—Pertaining to the skin.
Diaphanous—Translucent.
Digitigrade—Walking on the toes, like a dog.
Dorsal—Pertaining to the back.

Dorsal fin—The fin on the back of fishes.

Emarginate—Slightly forked or notched at the tip, or sometimes abruptly narrowed (said of quills.)

Endoskeleton—The skeleton proper—the inner bony framework of the body.

Epignathous—Having the bill hooked.

Erectile—Susceptible of being raised or erected.

Even—(*Tail*) having all the feathers of equal length.

Exoskeleton—Hard parts on the surface of the body.

Exserted—Projecting beyond the general level.

Facial—Pertaining to the face.

Falcate—Scythe-shaped; long, narrow, and curved.

Falciform—Curved, like a scythe.

Fasciated—With broad colored bands.

Fauna—The animals inhabiting any region, taken collectively.

Ferrugineous—Rusty red.

Fibula—The small outer leg bone.

Filament—Any slender or thread-like structure.

Filiform—Thread-form.

Fissirostral—Having the bill very deeply cleft, beyond the base of the horny part, as in the Swallows.

Forehead—Frontal curve of head.

Foramen—A hole or opening.

Forficate—Deeply forked.

Fossæ—The grooves in which the nostrils of many birds open.

Fossorial—Adapted for digging.

Fulcra—Rudimentary spine-like scales extending up the fins of some fishes.

Fuliginous—Sooty or smoky brown.

Furcate—Forked.

Fuscous—Dark brown.

Fusiform—Spindle-shaped; tapering toward both ends but rather more abruptly forward.

Gape—Opening of the mouth.

Ganoid—Scales or plates of bone covered by enamel.

Gastrosteges—Band-like plates along the belly of a serpent.

Gills—Organs for breathing the air contained in water.

Gill openings—Openings leading to or from the branchiæ.

Gill rakers—A series of structures like comb-teeth in the mouth of some fishes.

Glabrous—Smooth.

Gonys—The middle line of the lower mandible.

Gorget—Throat patch of peculiar feathers.

Graduated—Said of a bird's tail when the outer feathers are regularly shorter.

Granulate—Rough with small prominences.

Gular—Pertaining to the *gula*, or upper fore-neck.

Guttate—With rounded, drop-shaped spots.

Hallux—The great toe—in birds, the hind toe.

Height—Vertical diameter—practically the same as depth.

Heterocercal—Said of the tail of a fish, when unequal—the back-bone evidently running into the upper lobe.

Hirsute—With shaggy hairs.

Homocercal—Said of the tail of a fish when not evidently unequal; the back-bone apparently stopping at the middle of the base of the caudal fin.

Humerus—Bone of the upper arm.

Hyoid—Pertaining to the tongue.

Hypognathous—Having the lower mandible longer than the upper, as in the Black Skimmer.

Imbricate—Overlapping, like shingles on a roof.

Imperforate—Not pierced through.

Inarticulate—Not jointed.

Incisors—The front or cutting teeth.

Interfemoral membrane—The membrane connecting the posterior limbs of a bat.

Intermaxillaries—The bones between the superior maxillaries, forming the middle of the front part of the upper jaw, in fishes: the premaxillaries.

Jugular—Pertaining to the lower throat—said of the ventral fins, when placed in advance of the attachment of the pectorals.

Keeled—See *Carinate.*

Labials—Plates forming the lip of a serpent.

Lamellæ—Plate-like processes inside of the bill of a duck.

Lamellate—Said of a bill provided with lamellæ, as in a duck.

Lateral—To or towards the side.

Lateral line—A series of muciferous tubes forming a *raised line* along the sides of a fish.

Laterally—Sidewise.

Lobate—Furnished with membranous flaps—said of the toes of birds.

Longitudinal—Running lengthwise.

Loral plate—Plate between eye and mouth of a serpent.

Lore—Space between eye and bill.

Mailed cheeks—Having the suborbital bone extending over the cheeks, articulating with the preopercle (cheeks not necessarily hard or bony); said of some fishes.

Mandible—Under jaw (or in birds, either jaw.)

Maxilla—Upper jaw.

Maxillaries—Outermost or hindmost bones of the upper jaw, in fishes.

Metacarpus—The hand proper, exclusive of the fingers.

Metatarsus—The foot proper. (See *Tarsus.*)

Molars—The grinding teeth; posterior teeth in the jaw.

Moniliform—Necklace-shaped—widened at regular intervals.

Monogamous—Pairing; said of birds.

Muciferous—Producing or containing mucus.

Nape—Upper part of neck, next to the occiput.

Nasal—Pertaining to the nostrils.

Neural—Pertaining to nerves.

Nictitating membrane—The third or inner eye-lid, of birds, etc.

Nuchal—Pertaining to the nape or *nucha.*

Obscure—Dark, scarcely visible.

Obsolete—Faintly marked; little evident.

Obtuse—Blunt.

Occipital—Pertaining to the occiput.

Occipital plates—Plates on the head of a serpent, behind the vertical plate.

Occiput—Back of the head.

Ocellate—With eye-like spots, generally roundish and with a lighter border.

Ocherous—Brownish yellow.

Oid (suffix)—Like—as *Percoid*, perch-like.

Opercle, or operculum—Gill cover; the posterior membrane bone of the side of the head, in fishes.

Opercular bones—Membrane bones of the side of the head, in fishes.

Opercular flap—Prolongation of the upper posterior angle of the opercle, in Sun-Fishes, etc.

Opisthocœlian—Concave behind; said of vertebræ.

Orbicular—Nearly circular.

Orbit—Eye socket.

Oscine—Musical.

Oscine tarsus—By ellipsis, tarsus as in oscine birds; *i. e.*, its envelope undivided behind and forming a sharp ridge.

Oviparous—Producing eggs which are developed after exclusion from the body, as in all birds.

Ovoviviparous—Producing eggs which are hatched before exclusion, as in the Blind Fish and Garter Snake.

Palate—The roof of the mouth—in fishes, a part of the roof of the mouth, lying behind the vomer and in front of the pharyngeals (not to be confounded with either.)

Palatines—Bones of the palate.

Palmate—Web-footed, having the anterior toes full-webbed.

Papilla—A small, fleshy projection.

Papillose—Covered with papillæ.

Paragnathous—Having the two mandibles about equal in length.

Pectinate—Having teeth like a comb.

Pectoral—Pertaining to the breast.

Pectoral fins—The anterior or uppermost of the paired fins, in fishes, corresponding to the anterior limbs of the higher Vertebrates.

Pelage—The hair of a Mammal, taken collectively.

Pelagic—Living on or in the high seas.

Perforate—Pierced through; said of nostrils when without a septum.

Perissodactylous—Odd-toed (toes 1, 3, or 5.)

Peritoneum—The membrane lining the abdominal cavity.

Phalanges—Bones of the fingers and toes.

Pharyngeal bones—Bones at the beginning of the œsophagus of fishes, of various forms, almost always provided with teeth.

Pharyngognathous—Having the lower pharyngeal bones united.

Pigment—Coloring matter.

Plantigrade—Walking on the sole of the foot, as do men and bears.

Plastron—Lower shell of a turtle.

Plicate—Folded; showing transverse folds or wrinkles.

Plumage—The feathers of a bird, taken collectively.

Plumbeous—Lead-colored—dull bluish gray.

Pollex—Thumb; in birds, the digit which bears the alula—corresponding to the index finger.

Polygamous—Mating with more than one female.

Præcoces—Birds able to run about and feed themselves at birth.

Præcocial—Having the nature of *Præcoces*.

Premaxillaries—Same as intermaxillaries.

Premolars—The small grinders; the teeth between the canines and the true molars.

Preopercle—The membrane bone lying in front of the opercle and more or less nearly parallel with it; the "false gill covers."

Primary—Any one of the ten (often nine, rarely eleven) of the large, stiff quills growing upon the pinion or hand-bone of a bird, as distinguished from the secondaries, which grow upon the fore arm.

Primary Wing Coverts—The coverts overlying the bases of the primaries.

Projectile—Capable of being thrust forward.

Protractile—Capable of being thrust forward.

Pulmonary—Pertaining to the lungs.

Punctate—Dotted with points.

Pyloric cæca—Glandular appendages in the form of blind sacs opening into the alimentary canal of many fishes at the *pylorus* or passage from the stomach to the intestine.

Quadrate—Nearly square.

Quadrilocular—Four-chambered—said of the heart.

Quill—One of the stiff feathers of the wing or tail of a bird.

Quincunx—Set of five arranged alternately, thus
```
* *
 *
* *
```

Radius—Outer bone of fore arm.

Ray—One of the cartilaginous rods which support the membrane of the fin of a fish.

Rectrices—Quills of the tail of a bird.

Recurved—Curved upward.

Remiges—Quills of the wing of a bird.

Reticulate—Marked with a network of lines.

Retractile—Susceptible of being drawn inward, as a cat's claw.

Retrorse—Directed backward.

Rictal—Pertaining to the rictus, as *rictal* bristles.

Rictus—Gape of the mouth.

Rostral—Pertaining to the snout, as rostral plate.

Rudimentary—Undeveloped.

Ruff—A series of modified feathers.

Scansorial—Capable of climbing.

Scansorial tail—Tail feathers sharp and stiff, as in the scansorial birds (Woodpeckers).

Scapula—Shoulder blade.

Scutellate—Provided with scutella; said of the tarsus when covered with broad plates in a regular vertical series, and separated by regular lines of impression.

Scutellum—One of the tarsal plates or scutella.

Secondaries—The quills growing on the fore arm.

Second dorsal—The posterior or soft part of the dorsal fin, when the two parts are distinctly separated.

Sectorial tooth—One of the premolars of carnivora, adapted for cutting.

Semipalmate—Half-webbed; having the anterior toes more or less connected at base by a webbing which does not extend to the claws.

Septum—A thin partition.

Serrate—Notched, like a saw.

Sessile—Without a stem or peduncle.

Setaceous—Bristly.

Shoulder girdle—The bony girdle posterior to the head, in fishes, etc., to which the anterior limbs are attached.

Soft dorsal—The posterior part of the dorsal fin in fishes, when composed of soft rays.

Soft rays—Fin-rays which are branching and articulate.

Spine—Any sharp projecting point; in fishes, those fin-rays which are unbranched, inarticulate, and usually more or less stiffened.

Spinous—Stiff, or composed of spines.

Spinous dorsal—The anterior part of the dorsal fin in fishes, when composed of spinous rays.

Spiracles—Openings in the head or neck of some fishes and Batrachians.

Spurious—Said of the first primary when less than about one-third the length of the second. (The student will notice that in *Oscines* the *presence* of a short or spurious quill indicates *ten* primaries; its *absence*, *nine*.)

Sternum—The breast bone.

Striate—Striped or streaked.

Sub (in composition) — Less than; somewhat; not quite; under, etc.

Suffrago—Heel joint; tibio—tarsal joint.

Sub-caudal—Under the tail.

Sub-opercle—The bone immediately below the opercle.

Sub-orbital—Below the eye.

Subulate—Awl-shaped.

Superciliary—Pertaining to the region of the eyebrow.

Supra-orbital—Above the eye.

Syndactyle—Having two toes immovably united for some distance —as in the Kingfisher.

Synonym—A different word having the same or a similar meaning.

Tail—In mammals, the vertebræ, etc., posterior to the sacrum; in birds, the tail-feathers or rectrices, taken collectively; in serpents, the part of the body posterior to the vent; in fishes (usually), the part of the body posterior to the anal fin. (Everywhere used more or less vaguely.)

Tail Coverts—The small feathers overlapping the bases of the rectrices.

Tarso-metatarsus—The correct name for the so-called tarsus of birds; the bone reaching from the tibia to the toes, composed chiefly of the metatarsus, but having at its top one of the small tarsal bones confluent with it.

Tarsus—The ankle-bones collectively; in birds, commonly used for the shank-bone, lying between the tibia and the toes, the *tarso-metatarsus*.

Tectrices—The wing and tail coverts.

Temporal—Pertaining to the region of the temples.

Tenuirostral—Slender-billed.

Terete—Cylindrical and tapering.

Terminal—At the end.

Tertials—The quills attached to the humerus.

Tessellated—Marked with little checks or squares, like mosaic work.

Thoracic—Pertaining to the chest; ventral fins are thoracic when attached immediately below the pectorals, as in the perch.

Tibia—Shin-bone; inner bone of leg between knee and heel.

Tomium—Cutting edge of the bill.

Totipalmate—Having all *four* toes connected by webbing.

Tragus—The inner lobe of the ear; the lobe opposite the auricle.

Transverse—Crosswise.
Trenchant—Compressed to a sharp edge.
Truncate—Abrupt, as if cut squarely off.
Tubercle—A small excrescence, like a pimple.
Tympanum—Drum of the ear; external in some Batrachia.
Typical—Of a structure the most usual in a given group.
Ulna—The inner or posterior bone of the fore-arm.
Ungulate—Provided with hoofs.
Unguiculate—Provided with claws.
Unicolor—Of a single color.
Urosteges—The plates underneath the tail of a serpent.
Vent—The external opening ot the alimentary canal.
Ventral—Pertaining to the abdomen.
Ventral fins—The paired fins behind or below the pectoral fins in fishes, corresponding to the posterior limbs in the higher vertebrates.
Ventral plates—Gastrosteges in serpents.
Ventricle—One of the chambers of the heart.
Versatile—Capable of being turned either way.
Vertebra—One of the bones of the spine.
Vertical—Up and down.
Vertical fins—The fins on the median line of the body; the dorsal, anal and caudal fins.
Vertical plate—Central plate on the head of a serpent.
Villiform—Said of the teeth of fishes when slender and crowded into velvety bands.
Viscous—Slimy; viscid.
Vitta—A band of color.
Viviparous—Bringing forth living young.
Vomer—In fishes, the front part of the roof of the mouth; a bone lying immediately behind the premaxillaries.
Web—The vane of a feather, on either side of the rhachis or "stem"; also, the membrane connecting the toes.
Zygodactyle—Yoke-toed; having the toes in pairs—two in front, two behind.
Zygoma—The malar or cheek bone.

INDEX TO NAMES

OF

GENERA AND HIGHER GROUPS,

WITH THEIR DERIVATIONS.

NOTE.—In this index, names of genera recognized in this work are printed in ordinary type, as Dendrœca; families and higher groups in small capitals, as EMYDIDÆ; synonyms and sub-genera in italics, as *Amblodon*. In giving the etymology of terms, all words not otherwise designated are understood to be Greek; L. indicates Latin. Greek words are here, for convenience, printed in Roman characters.

www.ingramcontent.com/pod-product-compliance
Lightning Source LLC
LaVergne TN
LVHW010209110826
845151LV00002B/651

* 9 7 8 1 4 2 5 5 3 4 7 1 4 *